WITHDRAWN

ONCE AROUND THE GALAXY
Roy A. Gallant

FRANKLIN WATTS
NEW YORK/TORONTO/SYDNEY/LONDON/1983

Teaching Materials
523
Gall

Diagrams by the author

*Photographs courtesy of
The Lund Observatory: frontispiece;
Science Photo/Graphics Inc.: pp. 21, 22, 34;
Mount Wilson and Palomar Observatories: pp. 46
(top), 47, 52 (bottom), 58 (both), 65 (top);
Yerkes Observatory/University of Chicago: p.
46 (bottom); Hale Observatories: pp. 52
(top), 54, 55 (top right and bottom), 57 (both),
65 (bottom), 66 (all); Lick Observatory/
University of California: 55 (top left).*

*Cover photograph of the Great Nebula in Orion
courtesy of the U.S. Naval Observatory and NASA.*

*Library of Congress Cataloging in Publication Data
Gallant, Roy A.
Once around the galaxy.*

*Includes index.
Summary: A history of astronomy, focusing not
only on our current knowledge of the galaxy, but
also on how that knowledge has evolved over time.
1. Astronomy—Juvenile literature.
[1. Astronomy] I. Title.
QB46.G327   1983   523   83-10200
ISBN 0-531-04681-8*

*Copyright © 1983 by Roy A. Gallant
All rights reserved
Printed in the United States of America
5 4 3 2*

Contents

PREFACE	2
FOREWORD	3
1. FROM MYTH TO MEASUREMENT	5
Patterns in the Sky	5
Eudoxus and His Spheres	8
Aristotle and the Earth's Shape	10
Eratosthenes Measures the Earth's Size	10
Hipparchus Measures the Moon	11
The Zodiac	14
Precession of the Equinoxes	15
Ptolemy's Epicycles	19
2. OUR GALAXY GROWS LARGER	21
Copernicus Starts a Revolution	22
Tycho, the Master Measurer	26
Kepler, the Master Theorist	27
Galileo Builds a Telescope	29
Newton Writes a Scientific Law	30
3. A SHAPE FOR THE GALAXY	33
Halley: "The Stars Move"	33
Herschel: A Shape for the Universe of Stars	33
Distances to Nearby Stars	37
How Hot Are the Stars?	40
How Luminous Are the Stars?	42
How Massive Are the Stars?	43
Distance to the Distant Stars	45

4. A NEW SHAPE FOR THE GALAXY	47
A Matter of Scale	47
Shapley Measures the Galaxy	47
5. POPULATION OF THE GALAXY	51
The Galactic Nucleus	51
The Disk and Spiral Arms	55
Supernovae and Novae	61
The Milky Way's Corona	61
The Longest Year	63
6. BEYOND THE MILKY WAY	64
"Objects to Avoid"	64
The Universe Grows Larger	65
Galaxies Galore	66
Kinds of Galaxies	69
Galaxies on the Move	70
Help from Quasars?	72
7. ORIGIN AND FUTURE OF THE UNIVERSE	73
The Evolution of Galaxies	74
A Steady-State Universe?	74
The Big Bang	75
The Big Crunch	76
GLOSSARY	77
INDEX	84

FOR MARV AND ARLA

ACKNOWLEDGMENTS

The author wishes to thank the Board of Trustees at the University of Illinois for permission to include in this book several sections based on materials originally prepared for the University of Illinois Astronomy Program, on which the author served as a Science Education Specialist. These original materials are copyrighted © by the Board of Trustees, University of Illinois, although publication of the materials is not endorsed by the copyright holder. All sections of this book are based on popular courses in astronomy designed and taught by the author at the American Museum–Hayden Planetarium, New York City, and at the Southworth Planetarium, the University of Southern Maine.

Thanks also to Four Winds Press for permission to use brief passages from my books Fires in the Sky *and* Beyond Earth: The Search for Extraterrestrial Life, *and to Doubleday & Company, Inc., for permission to use selected brief passages from my book* Exploring the Universe.

Special thanks to Dr. K. L. Franklin, a longtime friend and colleague at the Hayden Planetarium, for his willingness to read and comment on the manuscript.

A view of our home galaxy, the Milky Way, as seen from the Earth and showing the location of several sky objects.

*Once Around
the Galaxy*

Preface

The late George Gamow, a physicist who delighted the many readers of his popular science books with a warm and witty style, liked to tell the following story about a renowned but fictional Russian philosopher named Prutkov:

"Which is more useful, the Sun or the Moon?" asks Prutkov. After some reflection, he answers himself: "The Moon is the more useful, since it gives us its light during the night, when it is dark, whereas the Sun shines only in the daytime, when it is light anyway."

This book is about a collection of more than 300 billion Suns and their planets and moons, enormous clouds of gas and dust, and other matter making up that gigantic island in the sky we call our home galaxy, the Milky Way. Many people are surprised to learn that the Sun is a star, and a rather ordinary star at that. It appears so much larger and brighter than the other stars we see only because we are so close to it. If we could push the Sun far out into space, among the other stars, it, too, would appear only as a faint pinpoint of light.

Many people are also surprised to learn that the Milky Way is the name of our entire galaxy, not just the name of the hazy band of light we see arcing up across the sky so prominently during midsummer and midwinter. Our knowledge of the Milky Way and of our place in it has grown slowly over many centuries. We have gained most of that knowledge in the past sixty years, and we have learned certain details about the shape of our galaxy only in the past twenty.

We will begin our account of how we have slowly pieced together our knowledge of the Milky Way by showing how people of long ago perceived the night sky, with its displays of meteors, auroras, comets, and other events regarded with fear and mystery. Along the way we will examine the sky through the eyes of astronomers who added important new bits of information that improved our understanding of the Milky Way—its shape, what it is made of, how it moves, and other of its properties. We will also try to answer questions such as, Are there other galaxies like our own in the Universe, or is ours unique? Are new stars now being born in the Milky Way? If so, how are they formed? How old is our galaxy, and what is its fate? Is it doomed to end as an unimaginably cold and dark place, its life snuffed out forever?

Come, follow me among the stars!

Foreword

Several years ago, I was urged by one of my students to design a popular astronomy course on the "dynamics of the galaxy," as he put it. "Exactly what do you have in mind?" I asked, a bit baffled, since the student in question had shown himself to be somewhat a of a variable performer, as stars go.

"I can tell you exactly what I have in mind," he said, carefully snapping open the rings of his notebook and presenting me with a neatly typed outline of what he wanted to know about our galaxy. Glancing down at the page, I was surprised and delighted to find that his interest had a dual focus. He was just as curious about how we have built our knowledge of the Milky Way as he was in acquiring a knowledge of the galaxy itself.

I have long been of the opinion that there is much to be gleaned from a study of the history of astronomy—that is, of the personalities who have held this or that notion, how they came to hold it, and their courage, or lack of it, in the face of new evidence. Hipparchus, Copernicus, Tycho, Kepler—all stubbornly clung to ingrained notions in spite of new evidence or convincing arguments to the contrary. The student I am speaking of has more intellectual comrades than we, as teachers, may be aware of—that is, individuals who have both the interest and ability to appreciate how certain ideas have evolved over time.

Accordingly, I designed a course about the "dynamics of the galaxy" within an historical context and have been teaching it for several years. This book is a by-product of that course, and I dedicate it to my students, past and present, who have become as excited over Kepler's and Newton's theories on why the planets stay in orbit around the Sun as over the current speculation on black holes or possible life forms on other planets.

Roy A. Gallant, Director
The Southworth Planetarium
The University of Southern Maine

February, 1983

1
From Myth to Measurement

In the beginning, says an Egyptian myth some 4,000 years old, there was nothing but a vast world-ocean, Nun. Then the first god, the Sun-god Atum, created himself out of the sea. When he rose out of the waters of chaos Atum said:

*Out of the abyss I came to be
But there was no place to stand.*

So he created a small mound of earth and stood on it, and it became the land. Then he created a god to rule over the sky, another to rule over the air, another to rule over moisture, and so on, while Atum himself ruled over all of creation.

A Chinese creation myth tells us that in the beginning there was only chaos. Then came the great creator-god P'an Ku. P'an Ku was a fearful creature with horns and fangs, and his body was covered with long hair. He began creating order out of the chaos by chiseling apart the formless world-matter and separating it into land and sky. He next sculpted the Earth's surface into mountains, valleys, and plains. He also created the Sun, Moon, and stars.

His work done, P'an Ku then died. His skull became the dome for the sky. His flesh became the Earth's soil. His bones turned into rocks. The rivers and seas formed from his blood. Trees and all other vegetation grew from his hair. The wind was his breath, the thunder his voice, the Moon his right eye, and the Sun his left. His saliva became rain.

The sky has long been a source of wonder, and often fear, to people of all lands. Comets, meteor showers, eclipses, and other unusual events had to be explained in some way. Not knowing the natural causes of these sky events, people of long ago invented supernatural causes. They inhabited the sky with gods, demons, and spirits, both evil and friendly. The sky became a vast battlefield where gods and demons fought in deadly combat and controlled the lives of mortals.

PATTERNS IN THE SKY

Long before writing was invented, people of ancient times populated the sky with fanciful pictures called the *constellations*. The Babylonians of the ancient Mideast, the Chinese, the ancient Maya of Central America, the American Indians, the Egyptians, the Arabs, the Greeks, and the Romans all invented constellations, although not necessarily the same or even similar ones.

The earliest known list of constellations was left by the Greek poet Aratus of Soli, who lived around 270 B.C. Aratus listed a total of 44 constellations, which he claimed to have gotten from the works of Eudoxus, who lived about a century earlier. Over the years, more and more constellations were added, with the result that today astronomers recognize a total of 88.

The constellations are an excellent example of how our senses can mislead us. This is especially true if we trust our senses without question when we try to estimate the distances and describe the motions of the stars, planets, and other objects in our galaxy.

When the ancient Greeks looked at the winter constellation Orion the Hunter, for example, they imagined all of his stars as being the same distance away from the Earth. For instance Rigel, marking Orion's right ankle in the illustration, Bellatrix, marking his right shoulder, and Betelgeuse, marking his left shoulder, were all thought to be the same distance from us. Even when we look at those stars through a large telescope, we cannot see how far away they are. Although Betelgeuse is the brightest *appearing* star in the constellation, it is not the brightest *shining* star. Rigel appears dimmer, but it is actually brighter.

Imagine for a moment that you are approaching a town by car at night. You see a long line of roadside lampposts stretching before you into the distance. The lamp right next to you appears brighter than the most distant lamp you can see down the line. But when you finally approach and pass that distant lamp, you see that it shines just as brightly as the first one you passed. And when you glance back at the first lamp, you find that it now appears dim.

Astronomers use the term *apparent brightness* to describe how bright a star or other light source appears to our eyes. They use the term *luminosity* to describe how bright the star, streetlamp, or other light source actually is. So we can say that although Betelgeuse has a greater apparent brightness, Rigel has the greater luminosity. Rigel is actually nine times more luminous than Betelgeuse. It appears dimmer because it is nearly twice as far from us.

And so it is with the other constellations. Their stars are arranged in space like leaves on a tree—some nearer and some farther away—not like raindrops on a window, all of which are the same distance from us.

So, Orion the Hunter, Taurus the Bull, Cancer the Crab, Leo the Lion, Pisces the Fishes, and all the other creatures of the celestial zoo are nothing more than imagined pictures traced by stars once thought to be all the same distance from us. Stargazers of old thought this because it is what their eyes told them. The stars appeared to glide across the surface of a great upside-down bowl, forming the heavens. Astronomers call this bowl the *celestial sphere*.

One example of how a constellation came to be named should be enough to show how difficult it is for us today to try to see the same shapes or figures in the sky that the ancients saw.

It takes a lot of imagination to see in any group of "fixed" stars the human or other figures once imagined by people to populate the night sky. An ancient representation of Orion the Hunter is shown here.

An Egyptian king who ruled about 235 B.C. had a wife named Berenice, who was known for the beauty of her hair. When the king went off to Syria to fight in a war, his wife promised that she would sacrifice her hair to the gods if her husband returned home unharmed. Later, on learning that he was indeed on his way home from battle unharmed, Berenice, true to her word, had her hair cut off. When the king saw her he became furious and stormed at his priests, demanding an explanation. The priests told him why Berenice had cut off her hair and said that it had been placed in the temple as an offering to the gods, who had watched over the king while he was in battle. They then reported that the hair had mysteriously disappeared during the night. It was a divine act, they said. So pleased were the gods with Berenice's sacrifice that they had placed her hair in the sky for all to admire. Thereafter, all the king had to do to see his wife's hair was to look up to the sky. Presumably, the king was pleased. The constellation Berenice's Hair is still with us today and can be detected in the spring sky just to the right of Bootes and above Virgo.

Although the tales of the Egyptians explaining the events in the sky can match those of the Greeks and the Babylonians in imagination, it was the early Greeks whose skill in geometry enabled certain relationships of motion among the Sun, Moon, planets, and constellations to be seen for the first time.

EUDOXUS AND HIS SPHERES

It was not easy for people of old to discard the ancient myths explaining the origin and plan of the Universe. The process began in Greece sometime around 600 B.C. The old superstitious beliefs rooted in polytheistic religions—those colorful tales of dragons, demons, and Sun-gods—came to be questioned and eventually were cast aside. People began to search for natural causes to explain what went on in the sky. Supernatural causes, such as a god driving a chariot across the sky to account for the rising and setting of the Sun each day, were no longer acceptable as explanations. That hazy band of light that the Greeks had named the Milky Way could no longer be explained away as milk spilled by the goddess Juno as she nursed the infant god Hercules.

One of the new thinkers was a Greek named Eudoxus, born in 408 B.C. Eudoxus looked for natural causes for the apparent motions of the stars, Sun, Moon, and planets. Each night the stars were seen to parade across the sky as a group, rising in the east and setting in the west. They never seemed to move in relation to each other, only as soldiers on parade. So these pinpoints of light came to be called the *fixed stars*.

But there were other sky objects whose apparent motions were different from those of the fixed stars. Nightly they could be seen to drift independently, sometimes eastward and sometimes westward, among the background stars. From time to time these mysterious objects would even appear to slow down, then back up and swing around in a loop before reversing direction again. Called *wandering stars* by the ancients, today we call these objects *planets* and know of nine. Until March 13, 1781, however, only five planets were known—Mercury, Venus, Mars,

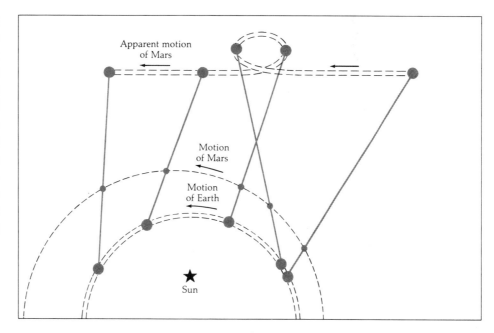

As we view the motion of Mars against the background stars, the planet appears to trace a looped path across the sky from time to time. The Earth's more rapid motion around the Sun causes this optical illusion

Jupiter, and Saturn. The ancients also looked on the Sun and the Moon as "planets," so to them there were seven in all. The Earth was not thought to be a planet. It was regarded as a very special place in the Universe, the very center of creation and the only abode of life. Furthermore, all the heavens and everything they contained appeared to move around the Earth while the Earth itself stood still, at the center of all there was. Or so it seemed.

Eudoxus' explanation of the "wandering stars" in relation to the fixed stars was an interesting one and was based on the apparent motions he observed, motions very different from the actual ones, which were not to be discovered for more than twenty centuries. Eudoxus imagined that the fixed stars were attached to a huge, invisible sphere of crystal and that the Earth was motionless at the very center of the sphere. He further imagined that this great star sphere slowly turned and so carried the fixed stars around, accounting for their rising in the east and setting in the west. To account for the motions of the Sun, Moon, and planets, he imagined a series of smaller spheres nestled within the outermost sphere, like the inner layers of an onion nestled within its outermost shell. While only one sphere was needed to account for the motion of the fixed stars, a set of five was needed for Saturn and each of the other planets.

These spheres of Eudoxus were a rather complex way to explain the apparent motions of the stars, planets, Sun, and Moon. But they did account rather well for those motions, although not quite well enough. For example, they failed to explain eclipses and the motions of Mars. The important thing was, however, that thinkers such as Eudoxus were trying to reason out a problem rather than blindly accepting the explanations of the ancient myths.

ARISTOTLE AND THE EARTH'S SHAPE

The great teacher Aristotle, who lived around 380 B.C., was another of the Greek thinkers who tried to solve the puzzles of the Universe by turning to reason rather than relying on the old myths. Although most people then thought that the Earth was flat, Aristotle reasoned that it could not be. The Greek mathematician Pythagoras had come to the same conclusion about 200 years earlier. You can see for yourself that the Earth is not flat, Aristotle explained, if you watch a ship sail over the horizon. First the hull slowly lowers from sight, then the mast. It is clear that the ship is not sinking at that time since a few days later it safely returns. Does this not prove that the oceans are curved and not flat? he asked. In addition, during an eclipse of the Moon, he said, the Earth's shadow cast on the Moon shows that the Earth is a sphere and not flat.

Aristotle's doubters—and there were many—could not understand why, if the Earth were a sphere, people on the underside did not fall off, or how they managed to walk "upside down." The law of gravitation, which would explain these things, was still 2,000 years in the future.

ERATOSTHENES MEASURES THE EARTH'S SIZE

At the time we are talking about, no one knew the Earth's size or the size of or distance to the Sun, Moon, or stars. Around 230 B.C., a librarian named Eratosthenes, who lived in Alexandria, Egypt, measured the Earth's size very accurately. He had dug a deep well in the town of Syene (now Aswan) in southern Egypt. On the first day of

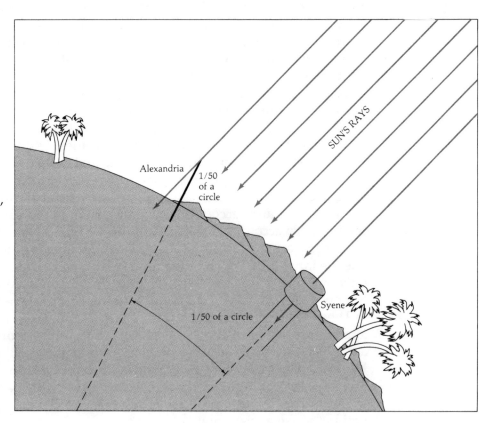

Around 230 B.C., the Greek scholar Eratosthenes very accurately measured the Earth's circumference by making observations of shadows cast by the Sun at two different locations on the Earth's surface.

summer he noticed that the Sun's rays shone directly down the well and were reflected off the water. But exactly one year later and 800 kilometers* to the north, in the city of Alexandria, he noticed that instead of shining straight down on top of a well house, the Sun at noon cast a shadow of 7 degrees 15 minutes.

Eratosthenes reasoned that since the Earth's surface was curved, the noon Sun must strike the two cities at two different angles. He also reasoned that the angle formed by the Sun's rays striking the well house in Alexandria would be the same as the angle formed by lines drawn from the Earth's center to each of the cities—a little more than 7 degrees, or about one-fiftieth of a circle. The final step was for Eratosthenes to multiply the distance between Alexandria and Syene by fifty. The result, he said, must be the Earth's circumference.

Eratosthenes was off by only about 200 kilometers, less than 1 percent. Today we know that the Earth's circumference is 39,800 kilometers. But even more remarkable to people of the time was the discovery that the Earth was so large—several hundred times larger than the landmasses then known.

HIPPARCHUS MEASURES THE MOON

The Moon's apparent motion against the fixed stars was much easier to understand than the planets' apparent motions. Each night the Moon could be seen to rise in the east and make its way westward across the great sky dome, following a fixed course through the same constellations month after month. It was also a simple matter for early stargazers to count the number of days it took the Moon to go through one complete phase change—from full moon to new moon and back to full moon again. The time taken is called a *synodical month* and is just a bit under thirty days.

Around 150 B.C., Hipparchus, called by some the greatest astronomer of ancient times, added several important facts to the growing body of knowledge about the heavens. First, he estimated the Moon's distance and size very accurately (to within 1 percent). His method was to measure the size of the Earth's shadow cast on the Moon during an eclipse. Today we know that the Moon's mean, or average, distance from the Earth is 384,400 kilometers and that its diameter is 3,476 kilometers.

The Sun's course across the sky dome was a bit harder to trace, since the stars were usually invisible by day. Even so, it could be traced accurately. Moments before sunrise, just before the stars fade from view, you can see part of a constellation behind the Sun. The particular constellation you see, of course, depends on the time of the year you are looking. At sunset, in the same way, you can see part of a different constellation forming the background of fixed stars. Again, the particular constellation you see will depend on what time of the year you are looking at the sky.

---

* Because U.S. astronomers and other U.S. scientists usually use the metric system for their calculations, we will use it here. A kilometer is approximately two-thirds of a mile (1 mile = .625 km).

In addition, there are times when we can see the stars very clearly during the day. During a total eclipse of the Sun, when the Moon blocks out the Sun's light, the stars gradually appear as more and more of the Sun's disk is hidden. At such times, the Sun's position against the background constellations can be seen.

By the time the Greek astronomers had begun a serious study of the stars, they had enough records of observations made by the Babylonians to chart the Sun's course across the celestial sphere without ever themselves having to observe the Sun at dawn or dusk. Their resulting view, or "model," as scientists say, of the celestial sphere showed very accurately just what constellations happened to be behind the Sun or Moon at any given moment of the day or night. In other words, once the Greeks had accumulated enough records of past observations—their own and those of the Babylonians—they no longer had to go outside and look to see where the Sun or Moon was in relation to a given constellation. Their tables told them, just as similar astronomical tables used by astronomers do today.

At the time Eratosthenes lived, another Greek scientist, Aristarchus, tried to calculate the Sun's distance by using the Moon's motion around the Earth. Although his method was good, his results were poor—some 130 times short of the Sun's actual mean distance of 149,600,000 kilometers. Hipparchus had also tried to calculate the Sun's distance, but his figure was ten times too small. His estimate of its size was also way off. Today we know that the Sun's diameter is 1,392,000 kilometers. Figuring out the Sun's distance was a problem for astronomers for many centuries. This was because the Sun was so far away that the angles used to determine that distance were extremely small and very hard to measure accurately.

What about the "wandering stars?" Did the paths they traced across the sky have any general motion in common? This was easy for the stargazers of old to figure out. During the nighttime, the paths followed by Mars, Jupiter, and Saturn across the constellations could easily be seen; and the paths followed by the remaining two planets recognized at the time—Mercury and Venus—were also known, since they followed along with the Sun. It turned out that the Sun, Moon, and five known planets all moved along the same path across the surface of the celestial sphere, passing through twelve constellations that formed a celestial highway, which came to be called the *zodiac*. The "center line" along that highway traveled by the Sun, Moon, and planets is called the *ecliptic*.

Before leaving Aristarchus, it should be mentioned that he was one of the first astronomers to regard the Earth as simply one of several planets, along with Mercury, Venus, and the others. He also believed that the Earth was not fixed and motionless at the center of the Universe but that, along with the other planets, it revolved around the Sun. In addition, the stars did not move across the heavens, Aristarchus said. Instead, the apparent nightly march of stars across the sky, and the Sun's apparent motion across the sky by day, were caused by the Earth's rotating on its axis.

This representation of the zodiac belt of twelve constellations is from Textus de sphaera, *published in 1531.*

Scholars of Aristarchus' time refused to accept the idea that the Earth moved around the Sun or that it rotated on its axis like a spinning top. Aristarchus could not prove that either idea was true. Also, the idea of a Universe with a motionless Earth at its center had been around too long to be upset so suddenly. Both of these important ideas were nearly 2,000 years ahead of their time.

THE ZODIAC **I**f we projected the Earth's equator onto the celestial sphere, we would produce the *celestial equator*. The zodiac is a belt of twelve constellations that appear to be wallpapered to the celestial sphere. As the diagram shows, the zodiac belt is tilted at an angle to the celestial equator, and the two cross each other in two places.

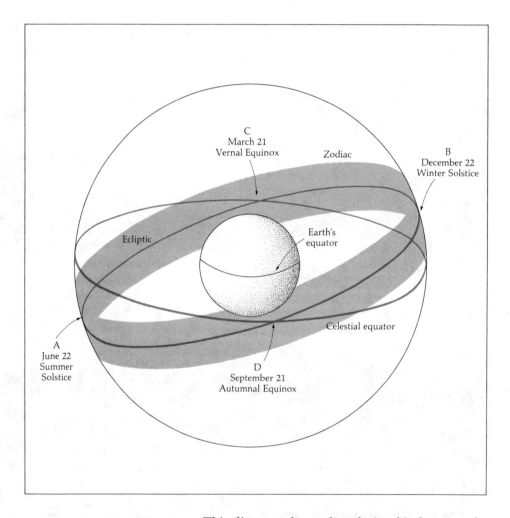

This diagram shows the relationship between the celestial equator and the ecliptic, which forms the center line of the zodiac belt. The Sun's apparent position along the ecliptic at A, B, C, and D marks the beginning of each of the four seasons.

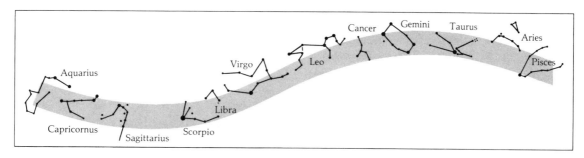

Stretched flat across the sky, this is how the zodiac band would look. It measures 360 degrees around the sky and is 16 degrees wide.

Each year the Sun and the planets appear to make one complete trip around the zodiac highway. This apparent motion is due to the Earth's orbital motion around the Sun and has nothing whatever to do with the Sun's real motion, which we will deal with in a later chapter. So we can picture the zodiac as an imaginary band 16 degrees wide and forming a complete circle of 360 degrees around the sky. Spaced unevenly along the zodiac are the twelve constellations. In order, they are: Aries the Ram; Taurus the Bull; Gemini the Twins; Cancer the Crab; Leo the Lion; Virgo the Virgin; Libra the Scales; Scorpio the Scorpion; Sagittarius the Archer; Capricorn the Sea-Goat; Aquarius the Water Carrier; and Pisces the Fishes.

PRECESSION OF THE EQUINOXES

Many centuries ago, the Sun's apparent path along the zodiac took it through the constellation Aries from March 21 to April 20. The Sun entered Taurus on April 21, left it on May 20, then entered Gemini on May 21, left it on June 21, and so on around the zodiac. Astrologers, who are people who tell fortunes by the stars, use these old dates even though they are no longer accurate. Today, the Sun enters Aries on April 18, Taurus on May 13, and Gemini on June 20.

The inventors of the zodiac several thousand years ago made an important observation over a period of many years. They saw that at the same time each year the Sun, in its apparent journey around the zodiac, crossed the celestial equator from south to north when it was in a center part of a certain constellation. They also made an important mistake. They supposed that this event always took place at the same time and in the same place each year.

At the time we are talking about, the Sun happened to be making this famous crossing about March 21, as it still does. We call that date the *vernal* (spring) *equinox*. There is also another equinox, the *autumnal* (autumn) *equinox*, which occurs on or near September 21. On this date the Sun appears to move down and cross the celestial equator from north to south as it continues its deceptive trip along the ecliptic.

You can imagine these two equinox points as opposite points on the celestial sphere. At each equinox there are very nearly the same number

—15

of hours of daylight and darkness, hence the name *equinox*, meaning "equal night."

At spring equinox, the Sun crosses the celestial equator from south to north, an event that marked the beginning of a new year, since it brought spring and signaled the start of a new planting season. Because the Sun entered Aries at this time of the year, it was natural to look on Aries as the first of the twelve constellations in the zodiacal zoo. The exact position in Aries marked by the Sun when it crossed the celestial equator came to be called the *First Point of Aries*. That term has survived to this day and is still used by astronomers and astrologers alike.

But Hipparchus upset this orderly picture of things. No, he said, the spring equinox had *not* always occurred at the same time. For a limited number of years it seemed to do so, but not over a long period of time. The actual dates for Aries—March 21 to April 20—were not fixed for all time but changed over the centuries. In fact, the spring equinox would never occur in any of the zodiac constellations for more than a certain length of time before moving westward into the neighboring constellation.

The ancient Egyptians, long before the time of Hipparchus, may also have been aware of this migration of the First Point of Aries along the zodiac. But Hipparchus seems to have been the first to work out the mathematics of it. He said that the equinoxes move westward along the zodiac at the rate of about 2 degrees of arc every 150 years. The spring equinox did occur in Aries at the time of Hipparchus, and it had occurred there since the year 1953 B.C. But by A.D. 220, which was 2,173 years later, the spring equinox was no longer occurring in Aries but in the neighboring constellation of Pisces, where it still occurs today.

So the First Point of Aries—that is, the spring equinox—crept its way along the ecliptic at a predictable rate. If you had lived in the year A.D. 220 and had been born on April 12, say, you would not have been born under the astrological sign of Aries. Instead, you would have been born under the sign of Pisces, as would anyone born on April 12 this year or next year. But from the point of view of astrologers, the dates for Aries will forever be from March 21 to April 20, regardless of how the stars actually move.

Although the spring equinox now occurs in Pisces, it will not remain there very much longer. In a little less than 400 years from now (about the year A.D. 2375), the spring equinox will have wandered westward out of Pisces and into the neighboring zodiac constellation Aquarius. Hence the "coming" of the Age of Aquarius.

This gradual apparent motion of the equinoxes around the zodiac is called the *precession of the equinoxes*, or simply the *precession*. Its cause remained a mystery until about 1687. At that time, the famous English physicist Isaac Newton framed his universal law of gravitation.

Today it is a simple matter to understand why the equinoxes precess, or creep around the zodiac. The Earth is not a perfect sphere. It rotates, and because of this it has developed a slight bulge around its equator. The diameter across the equator of our planet is some 42 kilometers more

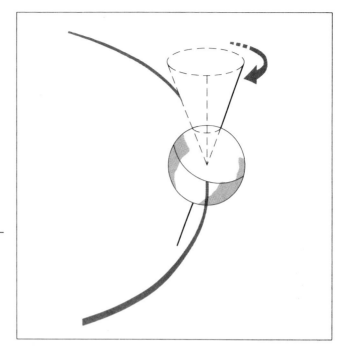

Because the Earth wobbles on its axis—"precesses"—there is a gradual change in the pattern of apparent motions of all the sky objects we can see.

than the diameter from the North Pole to the South Pole. Also, it happens that the Earth is tilted over at an angle of 23.5 degrees in relation to the plane formed by the ecliptic. Because of this lopsided arrangement, the Sun's and the Moon's gravitational tug on the Earth are not lined up and are forever twisting our planet this way and that. The result is that the tip of the Earth's axis wobbles around in a circle. One complete circle takes 25,800 years. This means that the North Star, presently Polaris, is not always the same star. At one time the polestar was Thuban, the third star from the end of the tail in the constellation Draco the Dragon. In a little more than 5,000 years from now Alderamin, the brightest star in the constellation Cepheus the King, will be the polestar. And about 7,000 years from now, Deneb, the brightest star in Cygnus the Swan, will be the polestar for a while, followed by Vega in the constellation Lyra the Harp in 13,000 years. There are long periods when there is no polestar at all. But for the time being, Polaris occupies the honored place, or very nearly so.

From about 600 B.C. up until the time Hipparchus died, which was around 120 B.C., the myths of old that had tried to explain everything that could be seen in the sky gradually lost their hold on the minds of thoughtful people. They were replaced—for a while at least—by a system of astronomy based on observation and measurement. Then the grand era of Greek learning came to an end when the Romans conquered Greece. Over the following centuries there was more emphasis on recording and preserving what Greek scholars had accomplished than on developing new ideas. So our knowledge of the universe of stars, much later to be recognized as merely one of countless "island universes," or galaxies, entered several centuries of a long, deep sleep.

This version of the Earth-centered universe as imagined by the Greeks shows the Earth at the center, then has symbols for the Moon, Mercury, Venus, the Sun, Mars, Jupiter, Saturn, and finally the enclosing belt of the zodiac. From Tabulae eclipsicum, *published in 1514.*

PTOLEMY'S EPICYCLES

One man who spent considerable time reviewing and summarizing all of the great astronomical work done before him was the Greek scholar Claudius Ptolemaeus, better known as Ptolemy. Ptolemy was also the last of the great original Greek thinkers. He lived around A.D. 130 and carefully recorded his many views on astronomy.

After long considering Aristarchus' idea of a Sun-centered Universe, Ptolemy rejected it. Instead, he gave his wholehearted support to the idea of an Earth-centered system, one in which the Earth stood still. He claimed—and rightly so at the time—that there were no observations to prove that the Earth moves around the Sun. Ptolemy also rejected the idea that the Earth rotates on its axis. If the Earth rotated, he reasoned, birds would have their perches whipped out from under them.

Ptolemy's plan of the Universe showed the Moon circling the Earth as its closest neighbor. Next came Mercury, also circling the Earth, then Venus, the Sun, Mars, Jupiter, and Saturn. Surrounding all was the crystal sphere of fixed stars marking the outer limits of the Universe.

Ptolemy did not completely accept Eudoxus' system of nestling spheres, however. Observation showed that Mars and Jupiter appeared brighter at certain times than at others. How could one account for the periodic brightening and dimming of these planets, especially as each traced out one of its occasional loop patterns? Ptolemy correctly reasoned that when bright, both planets must be closer to the Earth than when dim.

The theory Ptolemy favored to explain the plan of the Universe was one worked out earlier by Apollonius, a mathematician and astronomer who worked for a time with Eratosthenes in Alexandria. The theory had each planet moving about the center of a small circle while the center of the small circle moved around a larger circle enclosing the Earth. So, as a planet revolved around the Earth, it moved along in that well-known series of looped paths that came to be called *epicycles*. The larger circle, along which the epicycle loops were traced, was called the *deferent*.

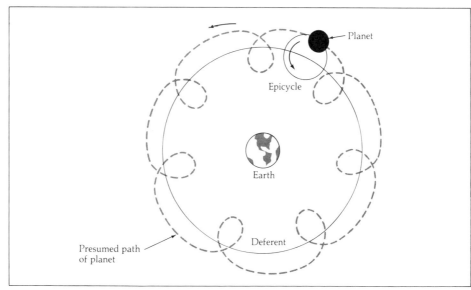

In the Earth-centered system of Ptolemy, each planet supposedly revolved around the Earth in looped paths called epicycles. The larger circle along which the epicycle loops were traced was called the deferent.

The beauty of this brilliant theory was that no matter how complicated the apparent motion of a planet might be, the addition of more epicycles could account for it. At least Ptolemy's reasoning led him to think so. He further said that with the epicycle theory he could accurately predict the position of any planet months or years into the future. His was the first system able to predict the motions of the planets with reasonable accuracy.

As well as being a brilliant theorist, Ptolemy was also an excellent observer. Earlier, Hipparchus had drawn up the most complete star catalog ever made. It showed the positions and relative brightnesses of 850 stars. Ptolemy enlarged the catalog to include more than 1,000 stars. When Ptolemy died his ideas—complete with errors—lived on and were to rule astronomical thought for the next 1,500 years.

2
Our Galaxy Grows Larger

We take up the story of our galaxy nearly a thousand years after Ptolemy had "explained" how the Universe was driven—as if by a mammoth machine whose gears meshed and turned and so moved the Sun, Moon, planets, and stars in their courses with the regularity of a flawless clock. After Ptolemy's death, and up to the 1500s, there were almost no new ideas in astronomy. Religious extremists forbade the kind of enlightened thinking that had made Greece great. Extremists among the early Christians and among the followers of Mohammed did their best to erase the work accomplished by Greek thinkers. They also did their best to discourage the quest for new knowledge about the planets and stars. For them, knowledge of God was the only important thing. If the great books of earlier scholars did not praise God's work, then they must be destroyed. And books praising God's work were simply not needed. So one after another the great libraries and centers of learning were burned. The only great books that survived were those hidden by scholars in private libraries.

In spite of this climate of gloom, among the many followers of Mohammed were a few who were determined to carry on the study of astronomy. One was the enlightened Arab ruler Caliph Harun-al-Rashid, who established a center of learning in the early 800s. Called the House of Wisdom, the center attracted many scholars, who brought with them treasured copies of centuries-old Greek writings. One such book was the work of Ptolemy, which Arab scholars called *Almagest*, meaning "the Greatest." All of these Greek works were translated into Arabic.

Al-Mamun, the caliph's son, also encouraged the scholars of his time to make and record observations of the sky and repeat the measurement of the Earth's size, as worked out earlier by Eratosthenes. The astrolabe and other fine instruments used by the Arabs of the time enabled them to make accurate measurements of the positions of the planets and stars.

By A.D. 1000, the Arabs had become very interested in the planets' motions and in Ptolemy's theory for predicting their future positions. However, when they plotted the positions of the planets on various nights and compared them with Ptolemy's predicted positions for those nights, they found that the two positions did not match. Ptolemy's theory

—21

had worked well enough over a period of a few years, but over long periods of time, his predictions had become more and more inaccurate. This had to mean that something was seriously wrong with Ptolemy's theory of epicycles.

Over the years, the Arabs put more and more emphasis on observational astronomy; this helped spark interest among European astronomers to reexamine the night sky. The Arabs' important role in keeping astronomy alive during that long period of intellectual drought is reflected by the large number of stars that were named for them, including Elnath, Zubenelgenubi, Alpheratz, Algol, Betelgeuse, and others, names that are still used by astronomers today.

COPERNICUS STARTS A REVOLUTION

Among the European scholars who were reading the old Greek masters was the Polish astronomer and Roman Catholic Church official Nicholas Koppernigk, known to us as Copernicus. Born in 1473, Copernicus was among those swept up in a new spirit of learning that had begun to spread over Europe.

Like Aristarchus and Ptolemy before him, Copernicus liked to build theories. He felt that the system of Aristarchus, which placed the Sun at the center of things and had the Earth and the other planets revolving around it, made more sense than Ptolemy's Earth-centered system. He also agreed with Aristarchus that the Earth's rotating on its axis caused the Sun, Moon, and stars to appear to parade across the sky from east to west.

It took Copernicus about thirty years to work out all the details of his *heliocentric* (meaning "Sun-centered") *system*. Yet even after he wrote his ideas down, he failed to realize the importance of them. If his friends had not urged him to publish them as a book, called *On the Revolutions of the Heavenly Spheres*, his work might have been lost to history.

It is said that Copernicus saw his book just after it had been printed and only a few hours before he died, in 1543. Apparently, he never thought it important—or wise—to communicate his theory to other scientists or to the public. Part of the reason, undoubtedly, was his fear that the Roman Catholic Church would punish him for daring to suggest that the Church-held view of Ptolemy's *geocentric* (meaning "Earth-centered") *system* could be wrong. Two other bold thinkers were soon to discover just how dangerous it could be to contradict official Church beliefs.

Copernicus' book was dull reading and sold very poorly. Yet it was the most important scientific work in more than a thousand years. It touched off the so-called Copernican Revolution and led scholars toward our modern view of the Solar System and its position within our home galaxy.

Copernicus had correctly said that the Earth spins on its axis, and he had correctly ordered the planets out as far as Saturn. But like Ptolemy he relied on epicycles to account for the puzzling back-and-forth motions

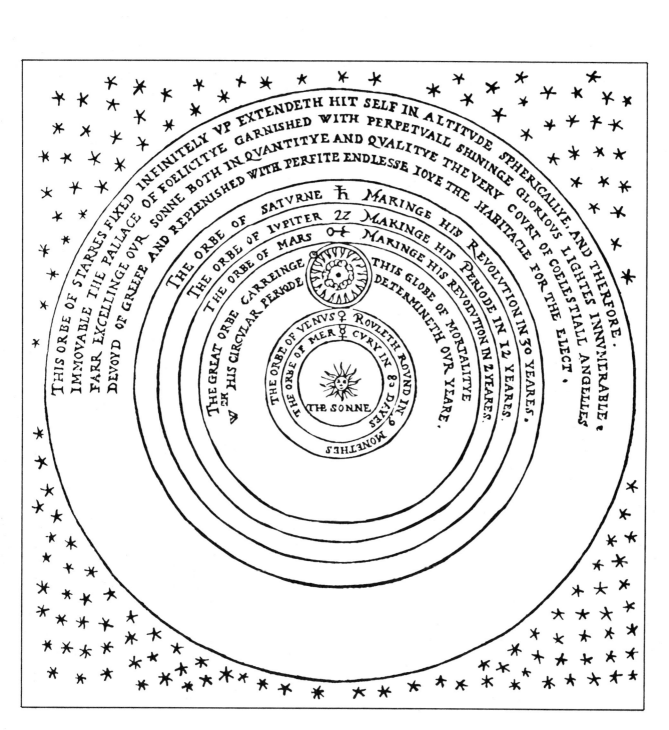

In the Sun-centered system of Copernicus, the Earth was a humble planet that revolved around the Sun along with the other planets. But like Ptolemy, Copernicus relied on epicycles to account for the looped paths that the planets seemed to trace.

By the 1600s, most astronomers no longer supposed that the Universe was contained within a great crystal sphere. They began to wonder what lay beyond the most distant stars they could see.

of the planets. He never realized that epicycles were not needed, for the looped paths seen to be traced out by the planets were nothing more than optical illusions. The illusion is caused by our observing the other planets from a planet that is itself moving around the Sun (see the diagram on page 23). To his dying day, Copernicus also believed that the distant stars were fixed to a great sphere of some sort, a sphere that marked the outer limit of the Universe.

Toward the end of the 1500s, several new ideas exploded into being, giving astronomers important new ways of thinking about the sky and the Solar System's place among the stars. Some of those ideas came from an English scientist named Thomas Digges. Digges correctly reasoned that: 1) It was the Earth's rotation that gave us the illusion that the stars paraded across the sky, but they actually did not move that way; 2) Since they didn't, then there was no reason to suppose that the stars were fixed to a great wheeling crystal sphere; and 3) Free of that sphere, the stars could lie at many different distances from the Earth—some relatively near and others very far away—at distances that we could not even imagine. And, he said, the stars were free to move in relation to each other. So the crystal spheres of Eudoxus, preserved and polished by Ptolemy, came crashing down.

Digges' reasoning began to convince astronomers that we live in a Universe infinitely large. Further, his teachings included a Sun-centered planetary system, with the Earth's position of importance reduced to that of a humble planet. It was Digges who so heavily influenced the thinking of England's scientists of the late 1500s.

One of the people influenced by Digges' teachings was a rebellious Dominican monk who was visiting England in the late 1500s named Giordano Bruno, born in 1548 near Naples, Italy. Bruno was the first of the two bold thinkers to feel the ruthless sting of the Roman Catholic Church. He recklessly criticized many Church teachings. As a result, he was expelled from the Dominican order and later excommunicated. Still, he continued to teach and to publish new ideas that clearly were not approved of by the Church. Among his writings was the following:

> *Sky, universe, all-embracing ether, and immeasurable space alive with movement...all these are of one nature. In space there are countless constellations, suns and planets; we see only the suns because they give light; the planets remain invisible, for they are small and dark. There are also numberless Earths circling around their suns, no worse and no less inhabited than this globe of ours. For no reasonable mind can assume that heavenly bodies which may be far more magnificent than ours would not bear upon them creatures similar or even superior to those upon our human Earth.*

Bruno also guessed that the Sun rotated, that some stars formed pairs, and that the Sun did not lie at the center of the Universe. He also supported Digges' idea of a Universe limitless in size. If the Universe had an edge, Bruno argued, then what lay beyond that edge? If the answer was "Nothing," then the world could be anywhere; and therefore the Universe could not have a center. With that argument Bruno removed the Earth—and humankind—from the center of creation. Although he had no proof to back up his ideas, he turned out to be right about most of them, including the Sun's rotation and double stars. Because of his outspokenness, however, and his harsh criticism of the Church, Bruno was arrested when he returned from England to his native Italy. He was tried and burned alive at the stake in the year 1600.

TYCHO, THE MASTER MEASURER

Astronomers knew that if a new theory did not fit the observed facts, then the theory should be thrown out, or at least changed. The data, or facts, could not be changed to fit the theory. Where was the evidence to support the heliocentric theory of Copernicus? It was to come from years of skywatching done by a fiery Dane named Tycho Brahe, born in 1546.

Tycho was not a theorist, but he was an expert observer. He never did accept Copernicus' heliocentric system, although the data he painstakingly collected and recorded over the years supported a Sun-centered system.

Tycho worked in perhaps the most unusual observatory in the history of astronomy. It was a gift, given to him by King Frederick of Denmark in the year 1576. Called Uraniborg, meaning "fortress of the heavens" (from Urania, the Greek muse of astronomy), the observatory was on the island of Hven. It consisted of a small castle, servants, a prison, and fine observing instruments. Tycho also received a handsome salary to help him carry out his "mathematical studies."

For the next twenty years, Tycho plotted the changing positions of the planets nightly. His instruments were the most precise ever made and he used them with a precision unequalled before. Tycho's measurements of planetary and star positions turned out to be five times more accurate than those of the great Hipparchus, even though telescopes had not yet been invented. Tycho's instruments required that the observer sight along them as you sight along the length of a rifle.

On November 11, 1572, Tycho observed a *supernova* in the constellation Cassiopeia. Rare events, supernovae were not understood in Tycho's time. Today we know that a supernova is a star whose outer layers of gases explode off into space. The star flares up and becomes many times brighter than before and may remain at that brightness for a few months. An earlier supernova had been observed and recorded by Chinese astronomers on July 4, 1054. The next one was to appear in 1604.

"Tycho's Star," as the sixteenth-century supernova came to be called, crumbled another old myth. Aristotle had said that the heavens were eternal and unchanging. Here was proof that they did change.

In November 1577, Tycho observed a comet with a bright head and long tail. He was able to plot its position night after night for more than two months. Eventually he came to realize that the comet lay a great distance away, off among the planets. It was not an event occurring only high up in the Earth's atmosphere, as Aristotle had supposed. Since the comet was farther away than the Moon and moved out among the planets, it would have had to have passed right through a number of those crystalline spheres along the surfaces of which the planets were supposed to glide. Here was proof that such spheres could not exist.

Tycho was deeply bothered by the idea of a Sun-centered system. He maintained that there was no observational proof for it. His senses told him that the Earth was motionless, and his strong religious beliefs made it hard for him to admit that the Roman Catholic Church could be wrong in such an important matter. On the other hand, Tycho the mathematician appreciated the mathematical argument of Copernicus. It was far more convincing than Ptolemy's theories. What was he to do?

He ended up by compromising. He maintained that the Earth was motionless at the center of the Universe and that the Moon and Sun revolved around the Earth. But he admitted that it was possible that the other planets could be moving around the Sun.

The truth of the matter was not to be found in Tycho's or anyone else's sensory observations but in the very records Tycho had faithfully kept for twenty-one years. Although Tycho was never to find that truth, his records were destined to reveal it.

KEPLER, THE MASTER THEORIST

In the year 1600, a young mathematics teacher, age twenty-eight, became Tycho's assistant. By this time Tycho had fallen out of favor and had lost his observatory on Hven. Emperor Rudolph II had then befriended Tycho and provided him with a handsome income, a castle, and an observatory near Prague, Czechoslovakia. But after only a year of work at the new observatory, Tycho died, leaving his young assistant, Johann Kepler, with a rich harvest of observations.

A firm supporter of Copernican theory, Kepler nevertheless supposed that the Universe was snugly contained in a spherical form, the outer region of which consisted of a shell of stars about 3 kilometers thick. Kepler set himself the task of discovering the paths the planets followed in their orbits around the sun. He began by studying Tycho's detailed records of the changing positions of Mars throughout the year. His calculations revealed two things: 1) that Mars did not move about the Sun at a steady speed; and 2) that it also did not revolve around the Sun in a perfect circle. Since the time of the ancient Greeks, scholars had supposed that all celestial objects moved in circles, since the circle was considered the most "perfect" of all geometric forms.

After ten years of hard work and many false starts and errors, Kepler eventually worked out the laws that finally explained how the planets orbited the Sun. These laws became some of the most important scientific discoveries in the history of science:

KEPLER'S FIRST LAW. The orbits of the planets are special elongated circles called *ellipses*. The Sun is one of the two points (each called a *focus*; pl., *foci*) around which each orbital ellipse is formed.

—27

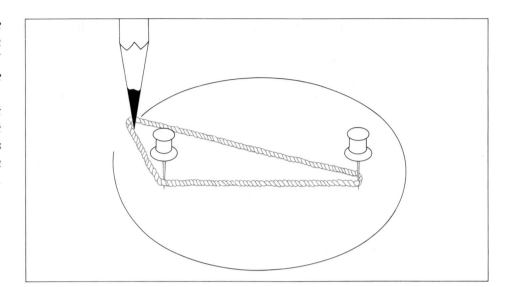

How an ellipse is drawn. Each pin is a "focus" about which the ellipse is formed. Kepler's first law said that the planets orbit the Sun in elliptical paths, not in circles.

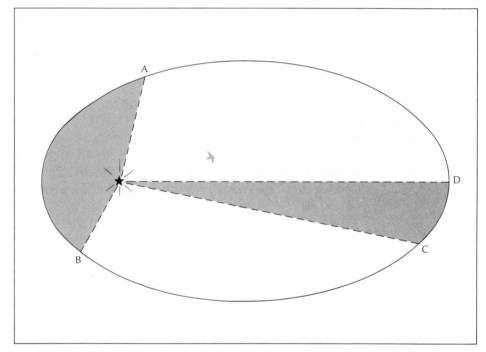

Kepler's second law said that a planet orbiting the Sun sweeps out equal areas in equal time. The two shaded areas are equal, as is a planet's orbital time from points A to B and C to D.

KEPLER'S SECOND LAW. A line joining a given planet to the Sun sweeps out equal areas in equal lengths of time. Thus, in the diagram, Planet P moves from Point 1 to Point 2 in the same amount of time it takes to move from Point 3 to Point 4.

Although Kepler labored hard to discover how the planets moved in their orbits, his efforts were never to provide an explanation of the force that kept them moving. This bothered him very much. He imagined mysterious "rays" of some sort sent out by the Sun as it spun around and whipped the planets along in their orbits. Kepler died in 1629, never having found the true explanation.

GALILEO BUILDS A TELESCOPE

The Italian astronomer Galileo Galilei, born in 1564, came close to providing an answer to the question of what drives the planets along in their orbits. Galileo, in 1609, was the first to use a telescope to study the heavens. The idea of a "spyglass" had been known by the Dutch and still earlier by the Italian Giovanni Battista della Porta. In 1610, Galileo's famous book *The Starry Messenger* announced to the world the wonders of the heavens revealed by his telescope. Where Copernicus' book had been dull reading, Galileo's captured the excitement and wonder of the stars, and the book appealed to scholars and the public alike. This was partly because Galileo had written his book in Italian, the language of the people, rather than in Latin, which was almost always used at that time for scholarly works.

Aristotle had taught that the heavenly bodies were spheres of perfection, smooth and unblemished. Galileo shattered this notion by describing craters, mountains, and valleys on the Moon. When he studied the Sun, he noticed blemishes there also—large, dark spots floating from west to east near the equator. Timing the spots as they crossed the solar surface, Galileo concluded that the Sun rotated once every twenty-seven days or so. Galileo's telescope also revealed four moons circling Jupiter, a tiny model of the Copernican system. He saw that Venus showed phases, like the Moon. And when he turned his telescope on the Milky Way, he saw that the hazy band was composed of countless stars. He wrote: "Upon whatever part of [the Milky Way] the telescope is directed, a vast crowd of stars is immediately presented to view. Many of them are rather large and quite bright, while the number of smaller ones is quite beyond calculation." There was no doubt in Galileo's mind that the stars were distant Suns and the planets visible to him were Earthlike objects.

Galileo reasoned that some force of attraction possessed by all the objects in the Universe kept the planets moving around the Sun. Although the general idea of gravitation occurred to him, he lacked the mathematical skills to frame a law of gravitation. This was not to be done for another seventy years.

Galileo also reasoned that objects in space did not have to be pushed along to be kept in motion. Once they were set in motion they simply kept moving, because space, he said, was frictionless. He explained this idea by asking his students to imagine a perfectly polished ship floating in a perfectly calm sea. Once pushed into motion, he said, the ship would continue to glide around the Earth forever, as long as there was no friction to slow it down.

Galileo's major goal was to convince the world of the truth of Copernicus' Sun-centered planetary system. Officials of the Roman Catholic Church warned Galileo not to teach this dangerous idea. The Church still supported Ptolemy's view of the Universe, and so its followers must do the same.

Galileo ignored the warning and, in 1632, published a book strongly praising Copernicus and making fun of Ptolemy. The pope became angry and turned the matter over to the Church court, the Roman Inquisition.

Galileo was arrested by the Church in 1633 and ordered to stand trial. During his trial, and kneeling before his judges, he was forced to deny that the Earth moved and "that the Sun is the center of the World." He was also forbidden ever to teach these ideas either in writing or in spoken word. He was then placed under house arrest and so lived out his remaining ten years. During that time his eyesight failed him, due to his observations of the Sun through his telescope. By 1637 he was blind. He died in 1642, but not before publishing his best scientific work, a book on mechanics and motion entitled *Two New Sciences*.

NEWTON WRITES A SCIENTIFIC LAW

Isaac Newton was born the year Galileo died. Newton, from Woolsthorpe, England, was to carry on where Galileo had left off. In 1665, London was struck by a deadly plague. Plagues had swept through Europe before and killed people by the hundreds of thousands. Shortly after the plague struck in 1665, Cambridge University shut down and sent its students home until the danger of the plague had passed. Newton, then age twenty-three, was one of those students. Interested in mathematics and the motions of the planets, he used his long vacation to begin thinking about what held the planets in orbit around the Sun.

Could it be, he wondered, that the force that caused a stone or other object tossed into the air to fall back to the ground was the same force that held the Moon in orbit around the Earth and the planets in orbit around the Sun? The more he worked on the problem the more he realized he could not solve it with ordinary mathematics, so he invented a new form of mathematics that came to be called calculus.

By the time Newton returned to Cambridge, in 1667, he had begun to think about gravitation, but he did not develop the roots of his idea until the end of 1684. Over the next twenty years, he drew up his three laws of motion, which solved the motion problems Galileo and the French philosopher Rene Descartes had unsuccessfully tried to solve earlier:

NEWTON'S FIRST LAW. A moving object will keep moving in the same direction and at the same speed until an outside force acts on the object. And an object at rest will remain at rest until an outside force acts on it.

NEWTON'S SECOND LAW. When a force acts on an object, the speed and direction of the object changes. The change in speed is in the same direction as the force. It is also proportional to the force but inversely proportional to the mass. In other words, it takes less force to move a light object than it does to move a heavy object.

NEWTON'S THIRD LAW. For every action there is an equal and opposite reaction. For instance, when you fire a shotgun you feel the gun "kick" against your shoulder. The kick is the reaction to the explosive action that pushes the pellets out of the gun.

We do not have to examine Newton's law of gravitation in very much detail to understand how it enabled astronomers to see how the Solar

System was held together and to calculate the masses of stars and planets. One principle of Newton's law of gravitation is that every object in the Universe exerts an attraction over every other object. Suppose that the Universe consisted of only two tennis balls. No matter how far away from each other they were, the tennis balls would attract each other, just as the Earth and Moon attract each other, and just as the Earth and Sun attract each other.

Newton also showed that the more massive two objects were, the greater the force of attraction between them. So two elephants in space would attract each other with more force than two walnuts. Newton further showed that the closer two objects were to each other, the greater the force of attraction. And he was able to show mathematically exactly how gravitational force changed with distance and mass.

Try to imagine the following situation, as shown in the diagram. It is gravitational attraction that keeps the Earth in orbit around the Sun. And the Earth's exact orbital path is dictated by the Sun's mass. In Part 2 of the diagram, notice that we have increased the Sun's mass. Also notice that because we have given the Sun more mass, the Earth's orbit is smaller and its orbital speed is greater. In Part 3 of the diagram, the Sun is less massive than in Part 1. Because the force of attraction is now less than it was at first, the Earth has been allowed to move outward and settle down into a larger orbit. It has also slowed down somewhat.

The Earth's distance from the Sun and its orbital speed would be different if the Sun's mass were either greater or smaller than at present.

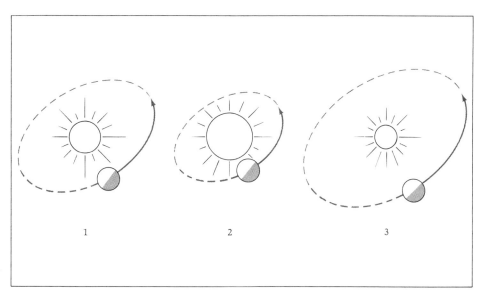

The planets, and artificial satellites orbiting the Earth, are all affected the same way by gravitation. For instance, notice in the following table how the orbital period and orbital velocity of an artificial satellite change as its distance from the Earth increases. (The orbital period of a planet or other body is the time it takes for that body to make one complete circuit in its orbital path around the Sun.) Also notice how the orbital period and velocity of a planet change as its distance from the Sun increases.

—31

ARTIFICIAL SATELLITES		
PERIOD (in hours)	DISTANCE FROM EARTH (km)	ORBITAL VELOCITY (km/hr)
2	1,688	25,319
7	12,207	16,676
12	20,240	13,933
17	27,195	12,406
21	32,237	11,562
24	35,871	11,059
PLANETS		
PERIOD (Earth days)	DISTANCE FROM SUN (km)	ORBITAL VELOCITY (km/hr)
EARTH: 365.26	149,500,000	108,000
MARS: 687.00	227,800,000	86,400
JUPITER: 4,331.98	778,000,000	46,800

Newton's universal law of gravitation beautifully explained the force that held the Solar System together, the force that neither Kepler nor Galileo had been able to understand. Newton wrote down all of his ideas about motion and gravitation and published them in a book usually referred to as the *Principia*, which appeared in 1687. To this day, the *Principia* remains one of the most important scientific works ever published.

Newton died in 1727. By that time people had come to accept Copernicus' Sun-centered Solar System and Digges' idea of a Universe infinite in size and populated with unimaginably large numbers of stars. The next problem was to find some order in the system of visible stars and to discover the place of our own star, the Sun, among those other stars.

3
A Shape for the Galaxy

HALLEY: "THE STARS MOVE"

Finding a shape for our galaxy was to involve the work of several astronomers and span more than two hundred years. Digges had suggested that the stars could lie at many different distances from the Earth and could move in relation to each other, although we could not see them do so in one lifetime. About a hundred years after Digges had died, the English astronomer Edmund Halley showed that the stars did indeed move. The year was 1718.

Halley is best known for the comet named after him, but his achievements in astronomy were many. At age sixteen he decided to become an astronomer, and by age twenty he was one. He was an excellent observer and plotted the positions of 360 stars from the island of St. Helena in the South Atlantic. When Halley returned to England in 1678, he began editing his material for publication. It later appeared under the title of *A Catalogue of the Southern Stars*, and it was this book that started Halley on his rise to fame.

In 1718, when Halley compared his star positions with those plotted by the Greek astronomers Hipparchus and Ptolemy about 1,500 years earlier, he was surprised to find that there were differences. For example, the bright stars Arcturus, Betelgeuse, and Sirius had all changed position over the centuries. Arcturus had shifted position by more than 1 degree.

Why don't we notice the actual, or proper, motions of the stars? A high-flying aircraft appears to move very slowly, even though it may be going more than 800 kilometers an hour. Another aircraft moving along at the same speed but at a lower altitude appears to be moving faster. The greater the distance of a moving object, the slower it appears to move, even though it may be speeding along at 100 kilometers a second. The stars are so very far away that astronomers need very powerful telescopes and many years to observe a star change position.

After Halley's discovery that certain stars move, astronomers began to ask if our own star, the Sun, also moved. Or was it instead fixed at the center of the Universe of stars?

HERSCHEL: A SHAPE FOR THE UNIVERSE OF STARS

If you look at the sky, and with your finger trace a small circle enclosing about twenty stars, you cannot detect any proper motion, no matter how hard or long you look at those stars. But if you look in a star catalog you can find the speed and direction in which each of the stars

—33

If you worked out the motions of stars seen in a given patch of sky, you would observe a helter-skelter pattern. Astronomers of the 1700s wondered if that apparent chaos of motion might, after all, be orderly.

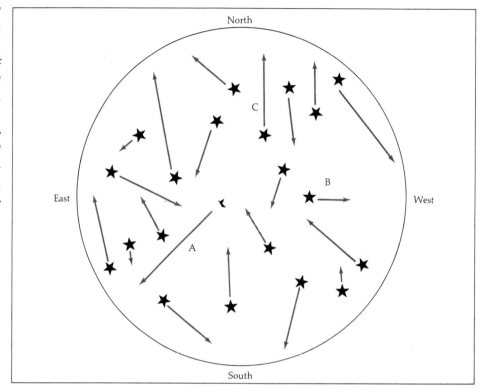

is moving. In the diagram, the directions the stars are moving in are shown by the arrow points, and the speeds are shown by the length of the arrows.

Any drawing you could make of your twenty stars would show a helter-skelter pattern. The stars appear to be moving in many different directions and at many different speeds. And your view from any other star in our galaxy would be the same.

On the night of March 13, 1781, the Englishman William Herschel, an organist by trade and an amateur astronomer, was observing the sky with an excellent reflecting telescope he had made. He was looking in the constellation of Gemini. Suddenly he noticed a strange "star." But it could not be a star, for stars appear only as pinpoints of light, due to their great distances from us. The object Herschel saw appeared as a disk, just as the planets do. Night after night he watched the mysterious object that ever so slowly glided eastward among the stars. Finally he decided that he had discovered a comet. He made detailed notes of his observations, which were sent to the Royal Observatory at Greenwich and to the university observatory at Oxford.

When other astronomers heard the news of a new "comet," they trained their telescopes on the object. After studying its motion, they found that it was following a near-circular orbit far beyond Saturn. The more they studied it, the less convinced they were that it was a comet. Finally they realized that what they were looking at was a new planet. Herschel's discovery doubled the size of the Solar System, for Uranus turned out to be twice Saturn's distance from the Sun.

Herschel decided to give up music and make astronomy a full-time career; two years later he made an even more important discovery. In 1783, he was able to announce that "the Sun is in motion through space." How did he arrive at this conclusion? Herschel had drawn direction and speed arrows for about a dozen stars. Although some of these arrows seemed to flare out from a point in the constellation Hercules, others seemed to close in toward a distant point in the opposite direction of the sky, a point in the constellation Columba. Herschel explained that this apparent flaring out of stars from Hercules and closing in of stars in Columba was an illusion caused by the Sun and its family of planets moving through space toward Hercules.

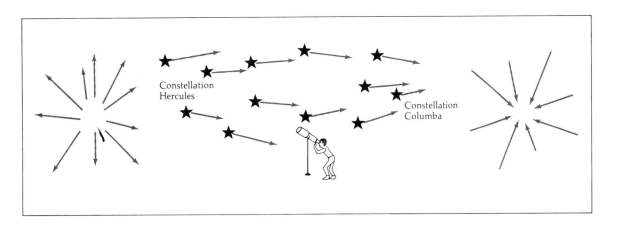

By observing its motion among certain other stars, Herschel found that the Sun moves through space. He noticed that stars in the direction of Hercules seemed to flare out of that constellation, while stars in the opposite direction, in the direction of Columba, seemed to converge. From this he concluded that the Sun was moving toward Hercules.

All of us have experienced the same illusion riding in a car. The objects far ahead all seem to be speeding toward us and flaring out from a point far ahead on the horizon. But when we look out through the rear window, all of the objects appear to be speeding away from us and closing in on a point far behind on the horizon. Herschel was right. Today we know that the Sun is speeding along toward Hercules and Lyra at 19.4 kilometers a second.

Another important discovery made by Herschel was the confirmation of Bruno's idea of double stars. In 1803, after several years of observing the changing position of a faint star associated with Castor, the brightest star in Gemini, Herschel was able to say that the fainter star was revolving around Castor in the same way that the Earth, for example, revolves around the Sun. Herschel had shown beyond doubt that Newton's law of gravitation works just as well out there among the distant stars as it does in the Solar System.

Herschel used his 18-inch (45-cm) telescope to make "star gauges," which enabled him to deduce the general shape of the Milky Way Galaxy.

Herschel was also the first to try to learn the shape of our stellar home. He began the search around 1800 and used his 18-inch (45-cm) reflecting telescope, which revealed about 10 million stars. Over several years he made a number of star counts, called *star gauges*. He would point his telescope first in one direction of the sky and count the stars visible to him in that patch of sky. Then he would point the telescope in another direction and make another count, and so on. While some patches of the sky contained many stars, other patches contained relatively few. Herschel reasoned that the patches crowded with stars had more stars extending out into space than did those patches containing fewer stars. Herschel made his star gauges from England. His son John later made additional gauges from the Southern Hemisphere. In all the two made about 6,000 gauges. By comparing them, Herschel concluded that our galaxy was shaped like a giant powder puff and that the Sun was about in the middle.

There matters stood for about a hundred years. The galaxy was a powder-puff array of stars, with the Sun in the middle, or so it seemed. It was the American astronomer Harlow Shapley who, in the early 1900s, was to give us our present view of the galaxy. But before he could do so other astronomers had to discover certain properties of stars and work out a way of measuring their distances from us.

DISTANCES TO NEARBY STARS

The next important breakthrough in probing ever deeper into the galaxy to learn about its size, shape, and composition came in 1838, when the German astronomer Friedrich Bessel worked out the first distance to a star other than the Sun. The star was one listed in the astronomers' catalog as 61 Cygni, a member of the constellation Cygnus the Swan.

As it turned out, two other astronomers had also calculated the distance to 61 Cygni, but Bessel usually is given the credit. The method he used, called the *parallax method*, is effective only for stars no more than about 500 light-years away. A *light-year* is the distance light travels in one year. Since light travels at the rate of about 300,000 kilometers per second, this comes to about 9.5 trillion kilometers.

You can see for yourself how astronomers use the parallax method to measure the distances to nearby stars by performing a little experiment and then following the simple arithmetic of solving a parallax problem.

Here's the experiment: Close one eye, make a fist, and hold it out at arm's length with your thumb sticking up. Next sight along your thumb to a tree, a telephone pole, or some other object. Now close the eye you have been sighting with and open the other one. Your thumb will appear to have changed position. If you alternately blink one eye and then the other, you will see your thumb jump back and forth. This apparent change in the position of your thumb is called *parallax shift*.

Bessel and others used the parallax shift of certain nearby stars to measure their distance from us, but they did not use their thumbs and blink their eyes. Instead, they carefully noted the position of a certain star in relation to the background stars. Then, six months later, when the Earth was on the opposite side of the Sun, they again sighted the star and compared its new position among the background stars to the old position. They next measured the distance between the two positions, called the *angular distance*. This gave them the number of degrees, or fractions of a degree, the star being measured appeared to move against the background stars.

Which star seems to have shifted position here in relation to its neighbors? The apparent shift of one object in relation to other objects, when viewed from two different positions, is called parallax shift.

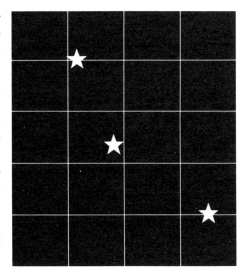
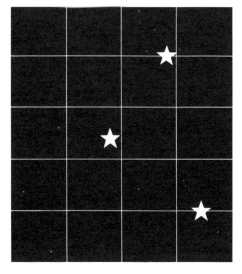

—37

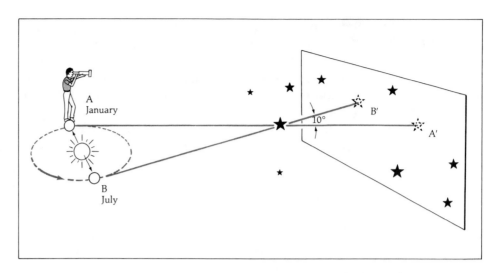

To measure the parallax shift of a star, astronomers photograph the star from one point in the Earth's orbit around the Sun, then six months later photograph the star from the opposite side of the orbit. In this example, a parallax shift of 10 degrees can be measured as the star seems to move from position A' to B' while the observer moves from position A to B.

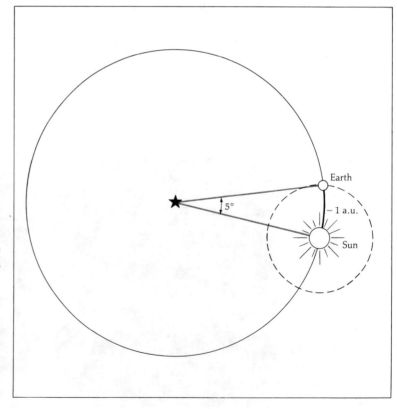

The parallax angle, or simply the parallax, is one-half the angle of parallax shift. In the example shown here, the parallax of the star being measured is 5 degrees.

In the diagram, notice that the top star has appeared to shift to the right by a certain amount. Now let's introduce a new term, the *angle of parallax shift*, and suppose that the angle of parallax shift is 10 degrees. Next we want the *parallax angle*, or simply the *parallax*, which is easy to calculate since it is half the angle of parallax shift, or, in our example, 5 degrees.

We are now well on our way to finding the distance to that star. We next draw a circle around the star, as shown in the diagram. Notice that the circle passes through both the Sun and Earth. The heavy line linking the Sun to the Earth marks off a distance that astronomers call an *astronomical unit* (a.u.), which is the mean (average) distance between the Sun and the Earth. Every 5 degrees around the edge of our circle marks off 1 a.u. Since a circle has 360 degrees, and since every 5 degrees make 1 a.u., the circumference, or distance around the circle, is 72 a.u.:

$$\frac{360}{5} = 72$$

Notice that the distance to the star is the radius, or one-half the diameter, of the circle. To find the radius all we have to do is find the diameter (d) and divide it in two. Once we know the circumference (c) of the circle, we can use the formula πd (or 3 × diameter) to find the diameter.

So,

$$c = 72 \text{ a.u.}$$
$$\pi = 3 \text{ (rounded off from 3.14)}$$
$$d = ?$$

$$\frac{c}{3} = d$$

$$\frac{72}{3} = 24$$

Since the diameter of the circle is 24 a.u., the radius is 12 a.u. And since the radius is the distance to the star, to find the radius in kilometers all we have to do is multiply the radius (12 a.u.) by the number of kilometers in 1 a.u. (150 million). That gives us 1,800,000,000 kilometers, or the distance to a star with a parallax of 5 degrees.

This may sound like a great distance, but if you look up the distance from Pluto to the Sun you will find that it is 6,000,000,000 kilometers. So a distance of 1,800,000,000 kilometers doesn't even get us out of the Solar System. The stars are much farther away from us than a mere 1,800,000,000 kilometers. Therefore, no star could possibly have a parallax of 5 degrees. The largest parallaxes measured for stars beyond the Solar System are only a small fraction of a degree. If you looked in a star catalog to find the distances to certain stars, you would come across a star called Barnard's Star. This is one of the nearest stars of all, but it has a parallax of only a tiny fraction of a degree:

—39

$$1 \text{ degree} = 60 \text{ minutes}$$
$$1 \text{ minute} = 60 \text{ seconds}$$
$$\text{parallax of Barnard's Star} = 0.5 \text{ second}$$

This means that 1 a.u. around a circle enclosing Barnard's Star is 0.5 seconds. One second is 2 a.u. along the circle. One minute would be 2 a.u. × 60 seconds, or 120 a.u. And 1 degree would be 120 a.u. × 60 minutes, or 7,200 a.u. The distance around that circle would be 360 degrees × 7,200 a.u., or about 2,600,000 a.u. Working as before, the radius of the circle, or the distance to Barnard's Star, turns out to be about 400,000 a.u., or 64 trillion kilometers. Using the much simpler units of light-years, the distance to Barnard's Star is 6 light-years.

The parallax Bessel measured for 61 Cygni was 0.35 seconds. That put the star about 9 light-years away. Since Bessel's time, this parallax has been corrected to 0.29 seconds, which means that 61 Cygni is 11 light-years away. To date, astronomers have measured the parallaxes for about 6,000 stars.

The farther away a star is, the smaller the parallax shift becomes. Therefore, there must be a distance beyond which parallax shifts do not occur and at which this method of measuring stellar distance fails us. That distance is about 500 light-years. To measure the distance of stars deeper into the galaxy, astronomers had to come up with a new method of measurement. But before they could do that, they had to discover certain other properties of stars.

HOW HOT ARE THE STARS?

If you have ever spent an hour or so looking at the stars on a clear night, you have probably noticed something about their color. Although many of them shine with a white light, some, such as Betelgeuse and Alpha Centauri, are reddish, while others, such as Capella and the Sun, are yellowish. Still others, such as Spica and Sirius, are bluish. A star's color is an important clue to its temperature and energy output.

Your everyday experience tells you that the temperature and color of objects being heated are related. For instance, a poker resting on glowing coals in a fireplace heats up until it becomes red-hot. If you blow air over the coals with a bellows, you can heat up the poker until it glows white-hot. If you could heat the poker still more, it would glow blue-white. This suggests that the bluish-white stars are the hottest ones, and so they are. The yellowish-white ones are less hot, and the reddish ones are the least hot.

This relationship between color and temperature provides astronomers with a valuable way of classifying stars and assigning temperatures to their surface gases. The Sun's surface gases have a temperature of about 6,000 kelvins, which is about 5,700 degrees C. Blue-white stars, such as Spica, Eta Aurigae, and Zeta Canis Majoris, are much hotter. They have surface temperatures of about 20,000 kelvins. Still other stars, including reddish ones such as Proxima Centauri and 61 Cygni A, are cooler than the Sun and have surface temperatures as low as 2,600 kelvins.

TEMPERATURE DEGREES CALLED KELVINS

Fahrenheit is not the only temperature scale. Scientists use two others, which are used in this book.

Most people of the world use thermometers marked in degrees C, for Celsius (also called centigrade). The C scale was invented by the Swedish astronomer Anders Celsius in the 1700s. As you can see in the diagram, the Celsius scale is the easiest one to use for everyday purposes. On it, ice melts at 0 degrees and water boils at 100 degrees, compared with the awkward numbers 32 and 212 on the Fahrenheit scale.

Objects get much colder than freezing, or 0 degrees C. But how much colder can an object get? Is there a limit? The British scientist William Thompson (also known as Lord Kelvin) thought there was. He supposed that if an object kept getting colder, the motion of its individual atoms would continue to slow down until they stopped. Atoms in a complete state of rest, he said, would have a temperature of *absolute zero*.

The temperature scale based on Kelvin's idea of absolute zero is called the *Kelvin scale* (also the absolute scale), and the degrees on that scale are called *kelvins*. Compare body temperature, room temperature, and the other temperatures shown on all three scales in the diagram.

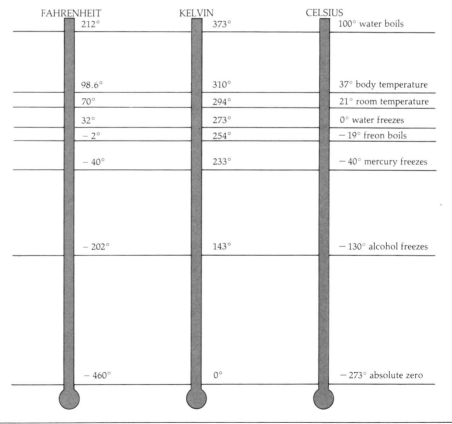

HOW LUMINOUS ARE THE STARS?

In the first chapter we said that astronomers use the term *luminosity* when they mean the actual brightness of a star, or the total amount of energy the star is producing. That energy includes visible light, heat, and all the other kinds of radiation given off by the star at any moment. Luminosity, you may recall, is not to be confused with how bright or dim a star appears to our eyes. Early in the 1900s, the Danish astronomer Ejnar Hertzsprung and the American astronomer Henry Norris Russell discovered an important rule that relates a star's temperature to its luminosity.

Look at the diagram below. Notice that we have assigned luminosity values up and down the left edge and temperature values along the bottom edge. Also notice the Sun's position on the diagram. If you read across from right to left, you will find that we have given the Sun a luminosity of 1. Reading down, you will find that the Sun's surface temperature is 6,000 kelvins. Notice, too, how the stars are arranged from the upper left to the lower right along the diagram.

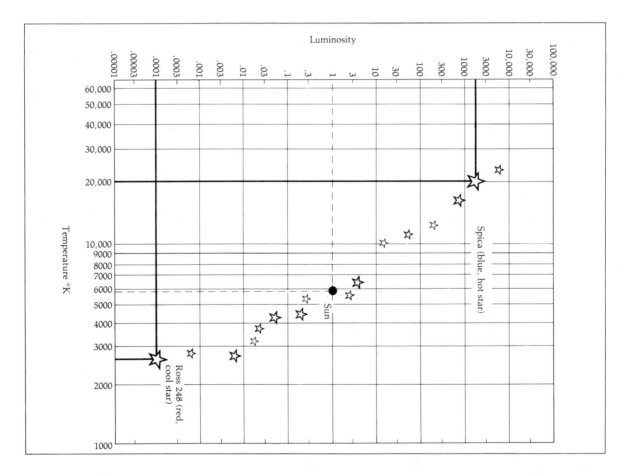

The Temperature-Luminosity diagram enables astronomers to estimate the temperature of a star when its luminosity is known, or the luminosity when its temperature is known.

Moving upward and to the left from the Sun, we find stars that are more luminous than the Sun and have higher temperatures. For example, the star Spica, up near the top, is 2,000 times more luminous than the Sun and has a surface temperature of 20,000 kelvins. This must also mean that Spica is much more massive than the Sun and much larger. It would take 2,000 stars like the Sun packed together to make one star as luminous as Spica.

Now look at the red dwarf star called Ross 248. It is only 1/10,000 as luminous as the Sun. It would take 10,000 stars like Ross 248 to shine with energy equal to that of the Sun.

So the color of a star is an excellent clue to the star's temperature. The star's temperature, in turn, is an excellent clue to the star's luminosity.

When Hertzsprung and Russell plotted many stars on a temperature-luminosity diagram, they found that most fell along a diagonal band that came to be called the *main sequence*. There are many dim stars that lie below the Sun and to the left of main sequence stars; and there are giant red stars that lie above the Sun and to the right of main sequence stars. But about 95 percent of all the stars astronomers have studied lie on the main sequence.

The temperature-luminosity diagram is a very useful tool. For example, when astronomers plotted on the diagram the positions of the hundred stars closest to the Earth, most of the stars fell below the Sun, among the red dwarfs. This revealed that most of the stars we can see are less luminous than the Sun. When they arranged the hundred brightest-appearing stars on the diagram, most fell above the Sun, along the main sequence. Only a few of these stars turned out to be nearby. So the temperature-luminosity diagram tells us that very luminous stars are rare in our neighborhood of space. Stars less luminous than the Sun seem to be the rule.

HOW MASSIVE ARE THE STARS?

You have seen that there is a large difference in luminosity among the stars. And there is a large difference in temperature. But what about mass? Do the stars differ as greatly in the amount of matter packed into them?

In the 1900s, the British astronomer Sir Arthur Eddington answered this question and gave astronomers important new information about the stars that populate our galaxy. He drew a "mass-luminosity" diagram that had luminosities listed along the left edge and masses listed along the lower edge. Eddington wanted to find out if there was a relationship between a star's luminosity and mass. If there was, he would then be able to tell from its luminosity how much mass a star had.

Notice in the mass-luminosity diagram that the Sun has again been assigned a luminosity value of 1. It has also been given a mass value of 1. This, of course, does not tell us anything about the actual luminosity or mass of the Sun. What it does tell us is how the Sun's mass compares with the masses of other stars more luminous than the Sun.

On his diagram Eddington positioned many stars whose luminosities and masses were already known. These were all double stars, whose

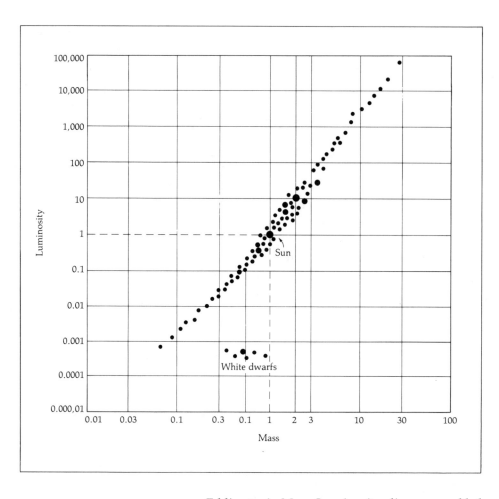

Eddington's Mass-Luminosity diagram enabled him to estimate the mass of a given star when its luminosity was known.

masses had been worked out by observing how they revolved around each other. Stars more luminous than the Sun were placed above the Sun's position. Stars less luminous were placed below. How massive a star was determined how much to the right or left of the Sun it would go in the diagram.

When Eddington had finished positioning many stars with known luminosities and masses on the diagram, he found a neat pattern. Nearly all the stars fell along a diagonal line. Since they did, he concluded that most stars whose masses were not known would also fall along that line. Say that a star is known to be ten times more luminous than the Sun. Its position on the diagonal would then be above the Sun on the diagonal and across from 10. To find that star's mass, all we have to do is read down to the bottom line, where the mass numbers are. By doing this we find that a star ten times more luminous than the Sun is twice as massive. And a star a thousand times more luminous is nearly ten times more massive.

Eddington was able to make up two general rules about the relationship between stellar masses and luminosities: (1) the more luminous a star is, the more mass it has; and (2) the less luminous a star is, the less mass it has. Because we know both the actual luminosity and mass of the Sun, the temperature-luminosity diagram and the mass-luminosity diagram can tell us the actual luminosity and mass of stars other than the Sun.

DISTANCE TO THE DISTANT STARS

You'll recall that the parallax method of measuring distance to the stars works well enough for stars that are closer than 500 light-years away from us. But what about stars lying at many times that distance? How do we find out how far away they are? The answer is in a law of light that describes what happens to the apparent brightness of a candle flame, a flashlight, or any other light source as it moves away from us, or we away from it.

Let's return to our earlier example of lampposts as we approach and pass them along a highway at night. As we close the distance between our car and a lamppost far ahead, the light appears to our eyes to grow brighter—and by a measurable amount. Because of the nature of light, when we close the distance by a half, we do not see the light twice as bright as before but *four* times as bright. When we then close the distance to one-third, we do not see the lights three times as bright as before but *nine* times as bright.

If we next watch out through the rear window as the lights fade off into the distance, we see them dim just as rapidly as they grew bright. When we have doubled the distance between our car and a receding lamppost, the apparent brightness does not dim by a half but by a fourth. Triple the distance, and the light appears not one-third as bright as before but one-ninth as bright. Because of the arithmatics involved, this law is called the *inverse square law of light*. (The force of gravitation, by the way, strengthens and weakens in the same way.)

Knowing that light behaves in this way, astronomers can attach a sensitive light meter to a telescope and measure how bright a certain star appears. If astronomers know the star's luminosity, they can then compute the star's distance by applying the inverse square law of light. But how?

Suppose that the telescope shows two stars whose color, and therefore temperature, are the same. It's a good bet that their luminosities will also be the same. But the apparent brightness of the two stars is very different. The star that appears brighter shows a parallax shift indicating that it is 65 light-years away. The other star does not show any parallax shift at all, which means that it is farther away than 500 light-years.

How can astronomers find the distance to the more distant star? By measuring its apparent brightness, scientists find the star to be only 1/81 as bright as the other star. Using the inverse square law of light, they can say that the dimmer star must then be nine times farther away than the brighter one. So the star with the weaker apparent brightness is 9×65 light-years away, or a little less than 600 light-years.

—45

With this method of measuring the distance to stars lying beyond 500 light-years, astronomers found an important new way to measure the size of our galaxy, detect its shape more accurately than Herschel had been able to, and accurately assign a position for the Sun within our local neighborhood of space.

4
A New Shape for the Galaxy

Have you ever tried to draw a scale model of the Solar System? If you have, maybe you were surprised to find that it is mostly empty space, with enormous distances separating the seemingly tiny planets.

A MATTER OF SCALE

Imagine the Sun as a pumpkin half a meter* across. On that scale Mercury, the nearest planet to the Sun, would be the size of a lettuce seed and would be 25 meters away. Venus, the second planet, would be the size of a pea and be 43 meters away. The Earth would be a slightly larger pea at a distance of 66 meters. Mars becomes a BB pellet 100 meters away from the pumpkin-Sun. The asteroids are the smallest imaginable grains of sand 183 meters away. Jupiter is an orange half a kilometer away, while Saturn is a lemon a little over three-quarters of a kilometer away. A little more than a kilometer away is Uranus, a large cherry, and Neptune, a somewhat smaller cherry, is 2 kilometers away. Finally comes Pluto, a small grain of sand 2.5 kilometers from the Sun.

If the distances between the planets seem vast on our pumpkin-Sun scale, the distances between the stars are much greater. Even if the stars were as small as pumpkins, those closest together would still be thousands of kilometers apart.

SHAPLEY MEASURES THE GALAXY

In the early 1900s, the American astronomer Harlow Shapley became the first to work out the size of our galaxy. The method he used was that of comparing the apparent brightness of certain stars with their luminosity, and so measuring the distance, as described in Chapter 3. Shapley and others were convinced that the center of our galaxy must lie off in the direction of the constellation Sagittarius, where the sky is crowded with stars. Off in the opposite direction, in the constellation Auriga, the sky has fewer stars, so we must be looking out of the galaxy when we look in Auriga's direction.

*A meter is approximately 3.28 feet, or a little over a yard (1.093).

Top left: *when we look off in the direction of Sagittarius, we find the sky crowded with stars, a clue that we are looking in the direction of the center of the Milky Way Galaxy.* Bottom left: *off in the opposite direction of the sky, we see far fewer stars, a clue that we are looking out of our galaxy into "empty" space.* Above: *globular clusters are enormous collections of about 100,000 stars, like this one seen in the constellation Hercules and designated M13.*

Between 1918 and 1921, Shapley studied the crowded section of the sky in Sagittarius and saw many huge collections of stars called *globular clusters*. A globular cluster is a spherical group containing more than 100,000 stars packed relatively close together. Of all the globular clusters Shapley could see, about 90 percent were in the direction of Sagittarius and only 10 percent off toward Auriga. Could it be, he wondered, that the globular clusters were arranged around the center of our galaxy like the fuzz of a dandelion around the center of its plant tip? If so, then the center of the galaxy would be the same as the center of the group of globular clusters. It seemed reasonable.

Suppose, Shapley reasoned next, that he could measure the distance to the center of the group of globular clusters. He would then have an estimate of the galaxy's size, since he would know the distance to its center. He was able to do this by measuring the apparent brightness of certain globular-cluster stars known as *RR Lyrae variable stars*. A variable star is one whose luminosity increases and then decreases in cycles. A typical RR Lyrae variable star builds in brightness for about five hours, until it reaches its brightest. Then it dims for about seven hours, until it reaches its dimmest, after which it begins to brighten again. One complete cycle, from brightest to dimmest and back to brightest again, takes about twelve hours for this type of variable star. That length of time is said to be the star's *period*.

RR Lyrae variable stars all have about the same average luminosity. Since Shapley knew the luminosity of the RR Lyrae variables in the globular clusters, and since he could measure their apparent brightness, he was able to estimate their distance by applying the inverse square law of light. Once he knew the distance to this or that RR Lyrae variable in this or that globular cluster, he knew the distance to the cluster itself.

By measuring the distance to many globular clusters, Shapley was able to estimate the distance to their central region, and so to the center of the galaxy. He calculated that distance at 100,000 light-years. So, according to Shapley, the distance from the Sun to the center of the galaxy was about 100,000 light-years. If the Sun were located out near the edge of the galaxy, then the distance to the opposite edge beyond the central region must be twice this, or 200,000 light-years.

Today's measurements show that our galaxy is only about half as large as Shapley thought. One cause of Shapley's error was fine cosmic dust spread among the stars. Astronomers of the early 1900s did not know of the existence of this dust, which was discovered in 1930 by Lick Observatory astronomer Robert J. Trumpler. Shapley had no way of knowing that the light from the RR Lyrae variables was being dimmed by dust as it crossed space to his telescope. The dimmer the starlight appeared, the more distant the star would seem to be. Even though the bothersome dust threw Shapley's measurements off, his work did much to help astronomers develop their present view of the Milky Way, a view that is changing today even as it did in Shapley's time and in the time of astronomers before him.

5
Population of the Galaxy

That luminous band of Milky Way stars, clearly visible in the summer and winter skies, divides the heavens into two hemispheres and marks the central plane of our galaxy. Binoculars clearly reveal individual stars and other features of the Milky Way band. If you scan back and forth along it, you cannot fail to see that the band widens and brightens noticeably in the direction of Sagittarius. Here lies the central hub of the galaxy, called the *nucleus*. We will begin our galactic census count in the nucleus.

THE GALACTIC NUCLEUS

The Sun lies some 30,000 light-years from the center of the galactic nucleus. Unfortunately, this enormous collection of stars is mostly hidden from view by dark clouds of gas and dust lying between us and it. We do know, however, that the nucleus is a slightly flattened sphere of stars, gas, and dust, with a diameter of about 33,000 light-years. It forms a gigantic central bulge similar to that seen in the Sombraro Galaxy and contains millions of densely packed red-giant stars that make it glow with a dull, reddish light. There is also a relatively thin scattering of gas and dust compared with the rest of the galaxy. It now seems that most of the dense clouds of gas and dust once in the nucleus were used up long ago during star formation when the galaxy was young.

Even though we can't see the individual stars of the nucleus, we know they are there. They reveal themselves by their strong outpourings of infrared, or heat, radiation. Gas and dust do not emit radiation along that part of the electromagnetic spectrum.

If we could visit the galactic nucleus and work our way into its center, we would pass through the onionlike layers of gas and dust that block our view of the central bulge. Just outside the nucleus and enclosing it is a ring of large clouds of hydrogen molecules, or pairs of hydrogen atoms. The outer edge of this ring is about 10,000 light-years from the center of the nucleus. The edge of the nucleus is marked by another ring, discovered in 1964 and also composed of hydrogen. The ring is rotating and is expanding at the rate of about 50 to 135 kilometers a second. It may be a spiral arm of matter being ejected out into the disk region of the galaxy. Or perhaps it is matter that was cast off by a huge explosion

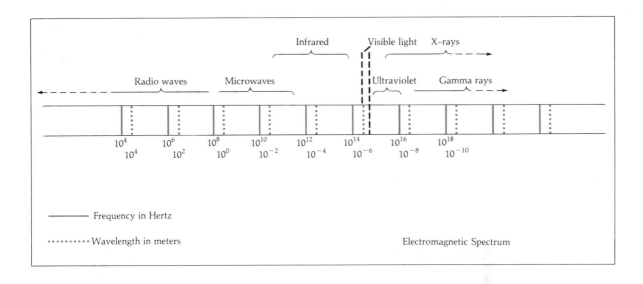
Electromagnetic Spectrum

millions of years ago. Such an explosion would have swept away the thin veils of gas and dust among the stars in the nucleus and so could account for the relative scarcity of gas and dust in the nucleus today. If we could sweep away the cloudy shells of gas enclosing the nucleus, we would see its many red-giant stars a trillion times brighter than we see them today.

During our journey to the galactic center we would come across a sprinkling of RR Lyrae variable stars, like those observed by Shapley in the hundred or so globular clusters that form a spherical halo around the galactic nucleus. The galactic halo has a diameter of about 130,000 light-years and contains a sprinkling of old stars plus about half of the galaxy's globular clusters. We would also come across many *planetary nebulae*, great shells of gas cast off by stars in their old age. Such stars have used up all their hydrogen fuel, and their nuclear furnaces have permanently shut down. They can no longer shine by fusing hydrogen nuclei into helium in their hot, core regions. Called *white dwarfs*, these dying stars have collapsed in on themselves and shrunk to densely packed objects about the size of Earth.

At a distance of about 1,000 light-years from the center of the nucleus, we would pass through another ring of large clouds made up of hydrogen molecules and charged hydrogen atoms. These clouds are hot spots, with temperatures of around 10,000 kelvins. They are associated with young and highly luminous, blue-white, supergiant stars. So along the way to the center, we would come across a sprinkling of blue-white supergiants, white dwarfs, and RR Lyrae variables among the overwhelming number of old red-giant stars.

About 35 light-years from the center, we would pass through still another ring of clouds. These are only 5,000 kelvins, and they revolve around the inner central region of the nucleus. About 10 light-years from the center, we would enter the most densely packed region of the galaxy and find ourselves surrounded by millions of red-giant stars whose distances from each other are less and less the closer to the center we go. There are also a scattering of dense clouds of charged hydrogen atoms in this region. An average cloud has about the mass of the Sun but a diameter of about 1 light-year. The stars and gas clouds in this part of the nucleus revolve around the central region once every 10,000 years or so. Clouds and stars still closer to the center revolve at a faster rate, just as the planets close to the Sun revolve more rapidly than the outer planets do.

The orbital motions of the gas clouds within the nucleus suggest that they may be satellites of some dense, supermassive object occupying the center. Astronomers know that something is there in the central region because it sends out strong infrared radiation and radio waves. The diameter of the object seems to be about 10 astronomical units, and its mass may be 50 million times that of the Sun. Some astronomers suspect that the object may be a *black hole*, a superdense mass that gravitationally pulls surrounding matter into itself. A black hole is so dense, and its gravity so strong, that theoretically nothing can escape from it, not even light. At this great density the star disappears inside itself, or within its so-called *event horizon*. After a dying star collapses as a black hole, it continues to contract until it is only a point, called a *singularity*.

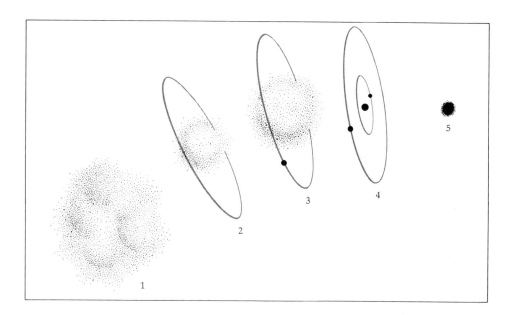

Stars form out of great clouds of gas and dust. The cloud contracts, heats up as a protostar, and one or more planets may form out of its disk matter (1). Later, the star shines as we see the Sun shining today (2). In old age, a Sunlike star exhausts its fuel supply and swells up, becoming a red giant (3). Eventually, the dying star contracts to a tiny object called a white dwarf (4). It then cools down and becomes a black dwarf (5).

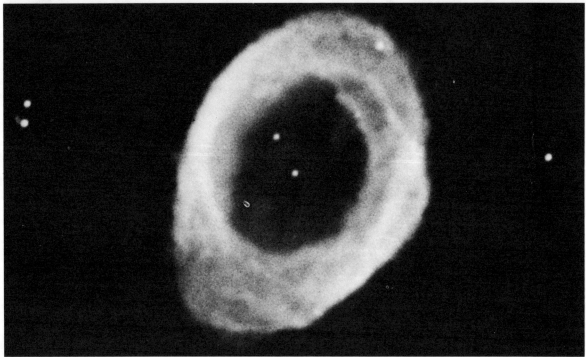

Above: *the Milky Way, a sprawling disk of stars with a giant spherical central region called the nucleus, is similar to this galaxy seen in Virgo and designated M104.*
Below: *the Ring Nebula in the constellation Lyra, designated M57, is a huge shell of gas cast off by an aging star.*

If not even light can escape from black holes, how do astronomers detect them? Gaseous matter pulled violently into a black hole from a nearby star seems to get heated to about 1 billion kelvins. Just before this matter disappears over the black hole's event horizon, it sends out strong bursts of X rays. Such X-ray bursts seem to be evidence of black holes. To date, the X-ray source known as Cygnus X-1 is the most promising candidate for a black hole, with a mass four to eight times that of the Sun.

Is there a black hole at the heart of our galaxy? If so, each year it could be gobbling up thousands of stars of the central nucleus, making this region of the Milky Way a stellar graveyard, one of explosive violence beyond our imaginations.

THE DISK AND SPIRAL ARMS

Our galaxy is a more or less flat spinning disk of matter spread out from the galactic nucleus in the shape of a thick phonograph record about 100,000 light-years in diameter. The so-called *galactic disk* may be only 1,000 to 6,000 light-years thick, or to scale, as thin as a dime or a quarter. That is not very impressive compared with the thickness of the nucleus. Even so, the volume of the galactic disk is some 10,000 times greater than that of the nucleus. The galactic disk also contains a much greater variety of matter.

In the 1940s, astronomers thought of our galaxy as containing a mass equal to about 40 billion Suns. By 1976, that estimate had soared to 900 billion Suns, and even that staggering amount seems low to some astronomers, who imagine a total galactic mass of about 1.2 trillion Suns!

Like certain other galaxies, the Milky Way is considered an open spiral galaxy because it has spiral arms wrapped around the nucleus. We know of three major spiral arms, which astronomers began tracing in the 1950s. Because many highly luminous, blue-white supergiant stars lie along the arms, astronomers have been able to trace the shape of the arms by plotting the positions of individual and clusters of blue-white supergiants. Many of these supergiants are surrounded by bright, glowing clouds of charged hydrogen, which can be seen in optical telescopes and detected with radio telescopes.

The three major arms of the Milky Way are: (1) the Orion Arm, named after the winter constellation Orion the Hunter and containing the Solar System; (2) the Perseus Arm, which curves farther out into the galactic disk than the Orion Arm by some 6,500 light-years; and (3) the Sagittarius Arm, which lies closer to the nucleus than we do by about 6,500 light-years. Recently, astronomers have detected an arm that seems to be an extension of the Sagittarius Arm. It joins the Sagittarius Arm in the Southern Hemisphere constellation Carina and so is called the Carina Arm.

Because the spiral arms of the galaxy are the only parts of the Milky Way rich with young, blue-white supergiant stars, they seem to be the principal places where star formation is taking place today. These highly luminous stars are each no more than about 10 million years old.

Above: *if we could view our home galaxy from afar, it would look to us very much like this beautiful spiral galaxy seen in the constellation of the Great Bear and designated M81.* Facing page, left: *astronomers can trace the pattern of the spiral arms in our home galaxy by detecting the bright glowing clouds of charged gases surrounding young, blue-white supergiant stars. These stars lie along the edges of the spiral arms. Among such stars and clusters of stars are the Pleiades, shown here.* Right: *the Horsehead Nebula, designated IC 434, is one of the most splendid dark nebulae visible to us. We see it outlined by light emitted from stars on its far side.* Below: *this is a composite photograph of the Milky Way as seen in the summer sky. The dark band, or "rift," is a dark nebula composed of dense concentrations of gas and dust.*

The inner edges of the spiral arms seem to be rimmed with dark clouds of densely packed gas and dust. These clouds are called *dark nebulae*; an example is the famed Horsehead Nebula seen in Orion. Another and more spectacular dark nebula is the dark band, or rift, running partway along the summer Milky Way.

The dark nebulae seem to be breeding grounds for stars. They are cold and do not emit light. We see the dark nebulae only when their shapes are silhouetted by rich fields of bright stars in the background. Because of the dimming and blurring effects of all this gas and dust, optical telescopes become useless to trace the shape of the spiral arms beyond a distance of about 20,000 light-years. When nebulae block the view through optical telescopes, astronomers must rely on radio telescopes to "see" beyond that distance.

Another way of learning more details about the structure of our galaxy's spiral arms and general shape is to study other galaxies that are very much like our own, such as the one shown on page 56. Telescopes useful for such study include large radio telescopes, such as ones in the Netherlands and in New Mexico, and the new optical Space Telescope scheduled to be placed in orbit by the space shuttle in 1985 or 1986.

The galactic disk contains nebulae other than the dark nebulae. Certain of those nebulae are called *reflection nebulae*, such as the gas and dust envelopes surrounding the stars of the Pleiades, shown on page 57. The gases in these nebulae do not emit light of their own but reflect light from the stars they surround, rather like a luminous fog surrounding a streetlight at night.

A third class of nebulae is represented by the Great Nebula in Orion, the most splendid of the so-called *emission nebulae*. The gas of an emission nebula emits a light of its own as it intercepts energy from nearby hot blue stars with surface temperatures greater than 25,000 kelvins. The surface temperatures of the blue-white stars embedded within a reflection nebula are somewhat lower. The gas and dust of emission and reflection nebulae are very thin. Their densities are far less than even the best vacuum we can produce in the laboratory.

Planetary nebulae are still another class of these objects. A famous one can be seen in the constellation Aquarius. It looks like a giant smoke ring floating in space. Actually, it is a huge shell of gas enclosing an intensely hot star.

Dying stars that cast off matter and produce planetary nebulae are the hottest stars we know of and have surface temperatures on the order of 100,000 kelvins. The gas they cast off reemits energy and glows with a fluorescent light at a temperature of about 10,000 kelvins.

The planetary emission nebulae were called "planetary" nebulae for two reasons. First, they were seen as a disk, like the planets; second, they were seen to glow with a pale greenish light and so were mistaken for planets when Herschel discovered them in the 1700s. There may be about 50,000 planetary nebulae in the galaxy, and they may form at the rate of about two a year.

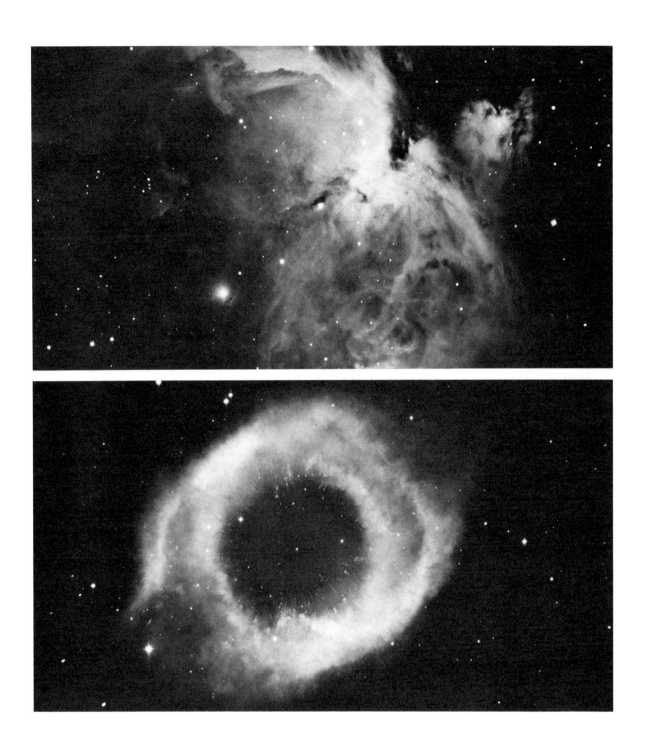

Above: *the Great Nebula in the constellation Orion is an example of an emission nebula, one whose gas reemits the energy of nearby hot-blue stars.* Below: *the planetary nebula in the constellation Aquarius, designated NGC 7293, is one of the most beautiful of these "shell-star" objects. They were once mistaken for planets because they appear as a small greenish disk of light.*

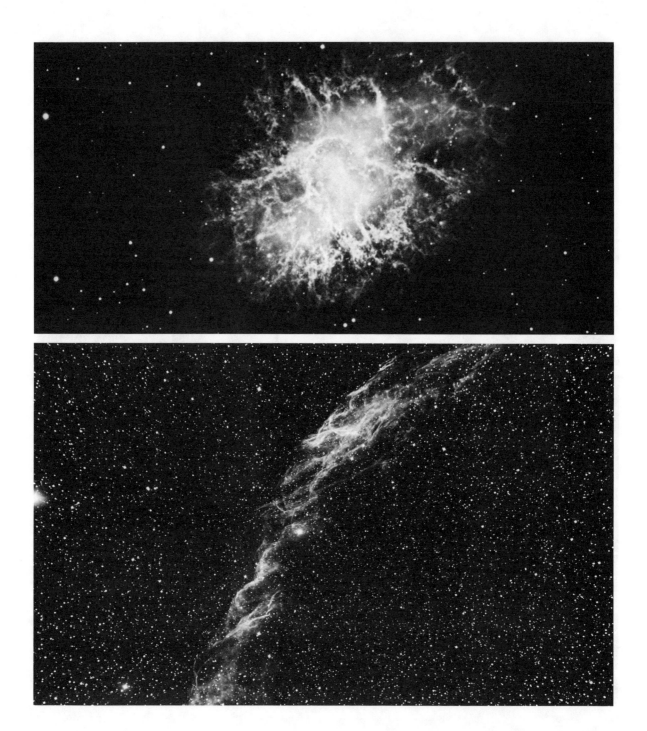

Above: *the Crab Nebula in the constellation Taurus, designated M1, is an expanding cloud of gas, the remains of a supernova observed by Chinese astronomers in the year 1054.* Below: *the Veil Nebula in the constellation Cygnus, designated NGC 6992, is the remains of a supernova explosion that took place thousands of years ago. Such supernova fragments cannot retain their identity for much more than 100,000 years.*

SUPERNOVAE AND NOVAE

Still another class of nebulae is caused by extremely high-energy stars that end their lives in catastrophic explosions—the supernovae, which were mentioned in Chapter 2. Two spectacular examples are the Crab Nebula in the constellation Taurus and the Veil Nebula.

Today we see the Crab Nebula as a cloud of matter about 4.2 light-years across and expanding at a rate of about 1,300 kilometers a second. Speeds that great can be triggered only by a gigantic explosion.

Only extremely massive stars can become supernovae. Stars like the Sun and the red dwarfs do not have enough mass to end their lives in a catastrophic explosion. A supernova is triggered when a giant star can no longer produce fusion reactions in its core. The star then collapses in on itself and explodes. The remains are a huge glowing cloud of matter surrounding the burned-out core, a superdense ball of neutrons called a *neutron star*. The diameter of such a stellar corpse may be only 16 kilometers or so, although the star itself may contain as much matter as the Sun, with a diameter 150 million times that of the neutron star. A lump of neutron starmatter the size of a sugar cube, if weighed on the Earth, would tip the scales at 10 million tons.

Some neutron stars spin wildly and send out rapid pulses of radio signals. These stars are called *pulsars* and were discovered by British astronomers in 1967. The neutron star at the center of the Crab Nebula is a pulsar rotating thirty times a second, one of the fastest pulsars found to date. An especially massive neutron star may eventually become a black hole.

Although the supernovae are the most spectacular stars to cast off matter into space, there are others. Stars called *novae* are lesser versions of the supernovae. They flare up suddenly and mysteriously, increasing in brightness over a period of days or several weeks. Then they return to normal again. Maybe fifty or so novae flare up in the galaxy every year.

So the galactic disk contains a wide variety of matter and many stars in a state of transition—novae, supernovae, nebulae, several kinds of variable stars, red-giant and supergiant stars, yellow, red, and white dwarf stars, the blue-white supergiants, and probably planets by the millions or billions. The stars of the galactic disk are younger than those found in the galactic nucleus and in the globular clusters.

THE MILKY WAY'S CORONA

As you have found in this book, and as you will find if you read other books touching on the history of astronomy, we have nearly always been a bit low in our estimates of the size or mass of this or that astronomical object. However, some of the numbers you read in this book will appear too large when compared with corresponding estimates in other books. For instance, the size of our galaxy is actually about 700,000 light-years, which is seven times larger than most quoted estimates of its diameter. That additional space is taken up by the *galactic corona*, a superhalo of matter, most of which is invisible even to present-day telescopes.

If we cannot see the galactic corona, so named by the Estonian astronomer J. Einasto, how do we know it is there? The answer lies in the speeds of the disk stars and the nebulae as they revolve around the galactic nucleus.

If the outermost edge of the galaxy were at its visible horizon, some 50,000 light-years from the center, then the velocities of stars there should be lower than the velocities of stars near the galactic center. After all, if Newton's law of gravitation works in the Solar System, shouldn't it work for the rest of the galaxy as well?

Those planets close to the Sun—Mercury and Venus—circle the Sun much faster than the more distant planets. Shouldn't we expect to find the same motion pattern out among the disk stars?

PLANET	AVERAGE SPEED AROUND SUN
Mercury	172,800 km per hr
Venus	126,400
Earth	107,200
Mars	86,700
Jupiter	47,000
Saturn	34,700
Uranus	24,500
Neptune	19,500
Pluto	17,600

Located about 30,000 light-years out among the disk stars, the Sun revolves around the hub of the galaxy at a speed of about 250 kilometers a second. Our Solar System chart of speeds tells us that stars farther away from the hub than the Sun should move progressively slower. But our measurements show just the opposite. Stars farther away than the Sun are revolving around the central hub faster than we are. Out at a distance of about 65,000 light-years, stars are speeding along in their orbits at 300 kilometers a second. This is 50 kilometers a second faster than the Sun. How can that be? Does Newton's law of gravitation break down once we leave the Solar System?

Fortunately it does not. Therefore, the only explanation is that there must be much more matter out there beyond the visible horizon of the galaxy. And that matter is gravitationally tugging at the stars and nebulae way out in the disk with a greater force than it tugs at the stars and nebulae closer to the nucleus. The result is that stars more distant from the Sun are speeded up in their orbits.

We can detect a number of objects farther away from us than 65,000 light-years. Among them are four globular clusters at around 130,000 light-years. Between 130,000 and 196,000 light-years are two more globular clusters and the Large Cloud of Magellan, which is a galaxy with a diameter of about one-third that of our galactic disk and visible to the naked eye only from the Southern Hemisphere. Between 196,000 and 260,000 light-years away are two dwarf, sphere-shaped galaxies plus the Small Cloud of Magellan, which is a companion galaxy of the Large Cloud and about two-thirds its size. Between 260,000 and 326,000 light-years away is another sphere-shaped dwarf galaxy plus three more globular clusters.

The American astronomer Bart J. Bok says that our own galaxy's surprisingly large size and mass, greater than believed only a few years ago, "has elevated the Milky Way to the rank of a major spiral galaxy." To account for the gravitational force acting on the outer disk matter, Bok feels that there must be hundreds of billions of solar masses lying undetected out there in the dark of the corona. It may be that much of that far-flung matter is too dim to be observed with present-day telescopes. Says Bok: "The best suggestion is that the corona of the Milky Way is composed mainly of old, burned-out stars. On the other hand, the unseen mass of the galaxy's corona may not fit any of the categories based on what can be seen in more accessible regions. We do not know yet what is out there."

THE LONGEST YEAR

As our galaxy spins around like a huge, rotating wheel, the stars and nebulae are carried around the nucleus. Stars and other matter at the Sun's distance from the galactic center are moving at a speed of about 250 kilometers a second. At that speed, how long does it take the Sun to make one complete trip around the galactic nucleus?

To find out the length of a so-called *cosmic* (or *galactic*) *year*, all we have to do is divide the circumference of the Sun's orbit around the galactic nucleus by the Sun's orbital velocity. A cosmic year is about 225 million Earth-years long. Half a cosmic year ago, the dinosaurs roamed across the American West, and the Earth was on the opposite side of the galaxy. Since our galaxy was formed, the Earth has made only about twenty trips around the galactic nucleus.

Think of the places we have been during that long time. In a sense, the Sun is a space probe that carries the Earth and all the other planets through the exotic forests of nebulae and stars that form the galactic disk. Since those objects are ever-changing, our view of the galaxy is never the same during any two trips around its center.

The question we must now ask is, What lies beyond the "edge" of the galactic corona?

6
Beyond the Milky Way

For centuries, astronomers thought of the vast collection of stars seen in the Milky Way as the entire Universe. They believed that beyond the limits of our star system there was nothing but the black, empty sea of outer space. If that view had turned out to be correct, our problems today in understanding the Universe might be much simpler than they actually are.

"OBJECTS TO AVOID"

For many years Galileo and astronomers after him peered through their telescopes in an attempt to understand the millions of pinpoints of light shimmering before their eyes. Occasionally, their view became blurred by mysterious clouds that seemed to exist in just about every region of space. Usually, they just ignored these clouds and turned their telescopes toward a clear patch of sky and continued their gazing.

One such astronomer was a French observer named Charles Messier. Messier's main interest in astronomy was watching comets. Some people like to collect shells or military buttons; Messier liked to collect comets. For years he scanned the heavens, but too often to suit him he would find his telescope pointed at misty patches of light in the sky, bothersome clouds of some sort. He took little interest in them, but, being a methodical observer, he made a record of each one he saw, listing them as "objects to avoid."

In the year 1784, Messier published a catalog listing 103 such objects. Each was given a number, such as M13. M13 is the globular cluster of stars visible in the constellation Hercules.

One of the most famous M objects is M31. On some clear and moonless night, when you are far away from city lights, look at the constellation Andromeda.

You can see it in the fall and winter sky. If you look carefully, you may see a faint, fuzzy patch of light across from the star Mirach. If you can't see it when you look directly at the spot, try looking a bit away so that you are looking out of the corner of your eye. The faint patch you see is the Andromeda Galaxy, listed by Messier as M31. This fuzzy patch of light was a riddle to astronomers for centuries.

—64

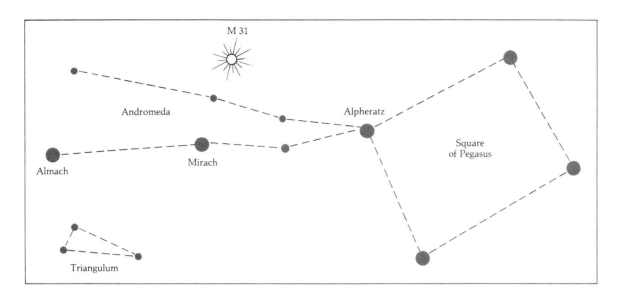

To find M31, the Andromeda Galaxy, first locate the Square of Pegasus, then the constellation Andromeda, which forms a crooked "V" to the upper left of Pegasus. M31 is just above the upper line of the "V."

THE UNIVERSE GROWS LARGER

Around 1791, Herschel suggested that the misty patches might be distant clusters of dim stars within the galaxy. But, as the years passed and Herschel studied the heavens more carefully, he began to wonder, "Is it possible that some of these star clusters are *not* part of the Milky Way? Could they be far out in space beyond the Milky Way? And if they are, does this mean there is more than one 'universe'?"

It wasn't until 1924 that these questions were answered. The man to provide the answers was the American astronomer Edwin Hubble. Working with the 100-inch (2.5-meter) telescope at the Mount Wilson Observatory, Hubble found a way of measuring the distance to the Great Nebula in Andromeda.

Photographs of the Great Nebula showed many individual points of light that appeared to be stars, but Hubble could not be sure that they were stars. For all he knew, they might not even be part of the nebula. But he decided to assume that they did belong to the nebula and that they were individual stars. It was a starting point.

As he studied the individual starlike objects, Hubble noticed that some of them changed from bright to dim to bright again, as did the RR Lyrae variable stars that Shapley had observed in the globular clusters. The pulsating stars that Hubble observed were also variable stars, but a superbright kind called *Cepheid variables*. There are many such stars in our own galaxy. They have periods of two to forty days and are about a hundred times more luminous than the RR Lyrae variables, which makes them 7,000 times more luminous than the Sun.

In the early 1900s astronomers had discovered a relationship between the length of the period and the luminosity of Cepheids. The longer the period of the star, the greater its luminosity. Measuring the periods of those faraway Cepheids in the Great Nebula told Hubble their luminosity. And knowing their luminosity enabled him to estimate their distance. All he had to do was measure their apparent brightness and apply the inverse square law of light. After making corrections for gas and dust between us and the Great Nebula, Hubble was able to estimate that the Great Nebula was about 1 million light-years away. But Hubble and other astronomers wanted more evidence for this great distance.

Nova stars helped provide such evidence. When novae in our home galaxy are at their brightest, they are about 100,000 times more luminous than the Sun. Several novae could be observed in the Great Nebula. Assuming that novae in the Great Nebula behave about the same as novae do in our own galaxy gave astronomers another standard of measurement. Since they knew a nova's luminosity when it peaked in brightness, they could measure its apparent brightness, apply the inverse square law of light, and then estimate the nova's distance. The distance turned out to be not 1 million light-years but 2. Which figure was correct?

Brightness measurements of globular clusters in Andromeda also gave a distance of about 2 million light-years, as did brightness measurements of extremely luminous, supergiant stars. The Great Nebula turned out to be a spiral galaxy similar to our own, but it may be about twice as large and have twice the number of stars. It is the only giant spiral galaxy close enough for us to examine in detail.

GALAXIES GALORE

With the discovery that the "nebula" in Andromeda was, in fact, a galaxy, the size of the Universe grew. We were no longer alone in space. And it continued to grow with the discovery that more and more of Messier's "bothersome" objects were galaxies, not patches of dust and gas. The number of "island universes" in the heavens is staggering to the imagination. A good telescope brings into view more galaxies than the number of stars seen by the eye. If you examined only the bowl section of the Big Dipper, you would find at least half a million galaxies! In all, astronomers think that about a billion galaxies can be seen through our largest telescopes. How many more lie beyond is anyone's guess.

In our local region of space, there are about twenty galaxies, making up what we call the *Local Group*. This includes three spiral galaxies, four irregularly shaped galaxies—among them the two Clouds of Magellan—and twelve others shaped like a slightly flattened sphere. Six members of this group are dwarf galaxies. The diameter of the Local Group is about 3 million light-years, which places the outer edge of our local cluster of galaxies three-quarters of the way to the Andromeda Galaxy.

Clusters of galaxies stretch off into space for as far as we can see with telescopes. Within 50 million light-years of us are many dozens of clusters similar in size to our Local Group. Clusters of galaxies form superclusters. One supercluster in the constellation Hercules is thought to be 350 million

Above: *the Andromeda Galaxy, M31, seen with its two companion satellite galaxies. Like the Milky Way, Andromeda is a spiral galaxy.* Below: *galaxies galore, including all shapes and sizes. This splendid cluster of galaxies is visible in the constellation Hercules. Don't expect to have such a grand view of the heavens with the unaided eye, though, or even with a small telescope. This photograph was taken through the 200-inch (5-m) Hale telescope.*

Above: NGC 205 is an elliptical galaxy and one of the two companions of M31 (see small photo, center). It is rich in red stars of the Population II class. The outer disk stars of Andromeda (see magnified section at left) contain many highly luminous giant and supergiant Population I stars. Left: like this one designated NGC 1300, some spiral galaxies have a straight bar running through them, with one spiral arm trailing from each end of the bar. Right: the so-called "peculiar" galaxies lack a recognizable shape, like this one listed as NGC 5128.

light-years across. Possibly the system of clusters and superclusters continues into super-superclusters. So far as we can tell now, our lumpy Universe of galaxy clusters stretches off in every direction of space for at least 10 billion light-years.

KINDS OF GALAXIES

Photographs taken with large telescopes reveal several thousand galaxies in enough detail that they can be classified, or grouped into different types. Instead of giving most galaxies names, astronomers assign them letter-number designations.

As we have seen, some galaxies have M numbers, such as M81. M81 is an *open spiral galaxy* (see page 56) with a bright central nucleus and spiral arms. M104 is a *closed spiral system* with a large nucleus. We see it nearly edge-on in the photograph on page 54. Surrounding the whole are many fuzzy specks. They form a great halo of globular clusters, like those surrounding the Milky Way.

Other galaxies have spiral arms extending from the ends of a straight bar instead of from a bright nucleus. NGC 1300 is a good example of a *barred spiral galaxy*. "NGC" stands for the *New General Catalogue*. NGC 205 is a fuzzy companion galaxy of the Andromeda Galaxy. It shows no sign of spiral arms or of a bright nucleus. Its outline is that of an ellipse, so it belongs to that class of galaxies called *elliptical galaxies*. There are also *irregular galaxies* that, like the two Clouds of Magellan, lack a recognizable shape; *spherical galaxies*, shaped like globes with fuzzy edges; and *transitional galaxies*, shaped like large cosmic footballs. There are even galaxies that do not belong to any of the categories already mentioned. These are called *peculiar galaxies*.

We can arrange galaxies in a sequence of forms, as shown in the diagram. The irregular galaxies, at one end of the sequence, are rich in young stars, like those in our own spiral arms. But they also contain old objects, such as RR Lyrae stars and globular clusters.

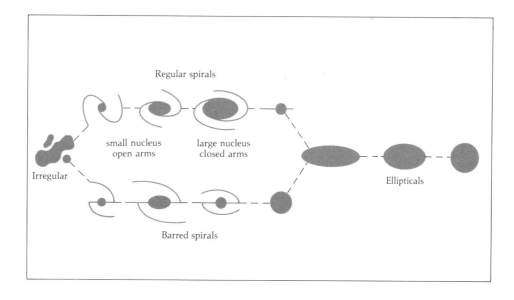

Next come the spiral galaxies, with small nuclei and loosely wound arms. These galaxies have fewer young blue supergiant stars.

As we move along the sequence from irregular galaxies to those with larger nuclei and more tightly wound spiral arms, the percentage of youthful stars dwindles. The percentage decreases even more as we proceed to galaxies like M104, which have even larger nuclei and more tightly wound spiral arms. At the far end of the sequence are the elliptical galaxies, which are made up almost entirely of old stars, like the oldest stars of our own galaxy.

Radio telescopes show that the amount of interstellar hydrogen decreases as we move in sequence from irregular galaxies to ellipticals. Perhaps up to half the matter in an irregular galaxy is interstellar hydrogen. In M51 it is only 5 or 10 percent; in our galaxy and in M31 it is only 1 or 2 percent. In elliptical galaxies there is scarcely any hydrogen.

The study of galaxies is so new that astronomers have not yet come to an agreement about how the shape of a galaxy might be related to its age—if there is any relation at all. Most astronomers think that all the galaxies may be about the same age, and that their different shapes come from the rotational speeds the galaxies were given at their birth and from the amount of dust and gas they contain. Those having a lot of dust have spiral arms. The others do not.

GALAXIES ON THE MOVE

By 1930 astronomers had studied enough galaxies to be sure of one thing. And though a few nearby galaxies were approaching us, all of the others were racing away at extremely high speeds. And though the Sun's speed around the Milky Way Galaxy was about 250 kilometers a second, many galaxies were racing away from our own at speeds of up to 15,000 kilometers a second!

Astronomers discovered something else, as well. The farther away from us a galaxy, or a cluster of galaxies, was, the faster it was speeding away from us. Listed on the next page are six galaxies and six clusters of galaxies. They are arranged in order of distance from us. As the distance becomes greater, what happens to the speed of the galaxy or cluster? No matter in what part of the sky we point our telescopes, the picture is the same. Galaxies and clusters of galaxies are rushing away from us at tremendous speeds.

To visualize what astronomers see, picture a half-inflated balloon with small paper dots stuck onto it, the dots representing the galaxies. As the balloon is inflated and expands, the distance between the dots becomes larger all around the balloon's surface.

Imagine yourself on one of these galaxy-dots, any one. As the balloon is inflated, every other galaxy-dot around you would appear to be moving away. From your position you would seem to be at the center of things, at the center of an expanding universe of galaxies. And if you moved to another galaxy, you would see exactly the same thing. No matter which galaxy you chose, the view would be the same. Every galaxy would appear to be at the center of a universe of galaxies expanding at breakneck speed.

GALAXY	SPEED (in km/sec)	DISTANCE (Millions of light-years)
M64	150	7
M82	300	13
M63	450	18
M65	805	25
M96	950	29
M60	1,110	38
CLUSTER	SPEED	DISTANCE
Coma	6,700	291
Pegasus II	12,714	490
Leo	19,500	847
Ursa Major 2	40,233	1,560
Hydra	60,600	2,600
3C 295	137,758	5,700

Let's stop and think about this for a moment. What does it mean? A very important law of physics tells us that the highest possible speed is the speed of light—300,000 kilometers a second. The law says that nothing can move faster than that. Now look again at the list of galaxies and clusters and read down the speed column. The highest speed we find is that for the cluster 3C 295. From what we can tell, it is moving at 137,758 kilometers a second—nearly half the speed of light. What if there are galaxies even more distant than 3C 295? If there are, we would expect them to be moving away from us at an even greater speed. And if there are galaxies still farther away, they must be moving even faster. Continuing in this way—if our thinking is correct—we would eventually be imagining galaxies moving at the speed of light—and faster! Here is one of the places where we begin to run into trouble.

Another trouble area is time. If we look at the Hydra cluster of galaxies tonight, we do not see that cluster as it is now. The light reaching us tonight from that distant cluster started out on its journey nearly 2 million years ago. And the light leaving that cluster right now will not reach us for nearly 2 million years. So when we look at the very distant galaxies and clusters we are seeing them as they were long, long ago. We have no way of knowing what they are like now or if they are still moving in the same direction that their light reaching us tonight tells us they were moving thousands and millions of years ago.

These are only two of the problems astronomers must face when they try to understand what the Universe is, what it was like in the distant past, and what it will be like in the far-distant future.

HELP FROM QUASARS? In the early 1960s, astronomers discovered a new class of objects in the heavens—the *quasars*. These mysterious points of light are the most distant objects known, the most luminous, and the speediest. They travel at more than 90 percent the speed of light—more than 270,000 kilometers a second. One of these objects releases enough energy in one second to supply all of the Earth's present electrical needs for a billion years.

The most distant quasars seem to be about 10 billion light-years away. So 10 billion years ago, when the quasars were emitting the light reaching us tonight, our galaxy was just being formed. If all galaxies are about the same age, and if the quasars are galaxies, it may be that we are looking back 10 billion years in time and seeing a galaxy being formed. Right now we simply do not know if this is the case. The puzzle of the origin of the galaxies—that is, of the Universe itself—is astronomy's superproblem.

7
Origin and Future of the Universe

Although astronomers disagree over many of the fine points of theories concerning what happens inside stars and how stars age, they are in general agreement about several larger ideas: (1) that stars are born out of the nebulae, shine for tens of millions to billions of years, and then die; (2) that the galaxies were born out of supernebulae and that they, too, after billions of years fade to dimness as their stars go out; (3) that the Universe appears to be expanding, meaning that the galaxies seem to be rushing away from us in all directions; and (4) that the galaxies we can see are distributed fairly evenly throughout space, although vast empty regions occur here and there. In general, there seem to be about as many galaxies off in any one direction of the sky as there are in any other direction. (There is an exception, though—a superhole in the Universe around the constellation Bootes and handle of the Big Dipper. The hole takes up about 1 percent of the entire sky.)

What all this means is another matter. In a way, we are in a fix similar to the one the ancient Greek astronomers were in, only ours is on a much larger scale. The Greeks could follow the paths traced by the planets across the "surface" of the sky, but they had difficulty envisioning those paths in three-dimensional space. Today we can detect certain motions of the galaxies—a general rushing away from us and from each other in three-dimensional space—but we cannot detect their sideways motion because of their great distances. We are unable to envision either a shape or a limit to the space occupied by the galaxies.

How do astronomers detect the motion of a galaxy away from us or toward us? The method they use is the same one used to detect a star's motion toward us or away from us. By attaching an instrument called a *spectroscope* to a telescope, an astronomer can obtain a spectral portrait of the star or galaxy, a kind of "fingerprint" of the light pattern emitted by the object. If the star or galaxy is rushing away from us, then certain dark lines in its spectral portrait are shifted toward the red end of the spectrum. If the star or galaxy is rushing toward us, then those lines are shifted toward the blue end of the spectrum. The greater the shift toward the red or blue, the greater the speed of the star or galaxy toward or away from us.

THE EVOLUTION OF GALAXIES

There are galaxies galore, and they come in a variety of shapes. What does this assortment tell astronomers about how a given galaxy has changed over the 10 billion or so years it has been around?

Not very much, because we haven't found enough clues to suggest how galaxies evolve after they form out of great turbulent clouds of gas. But, as with the clouds of gas and dust out of which stars form, we can imagine that some of the original galactic clouds were more massive than others. Otherwise, how can we account for some galaxies being dwarfs while others are giants? We can also imagine that some of the original galactic clouds were rotating rapidly, while others rotated hardly at all.

A cloud that has little rotation to begin with may collapse rather quickly, without flattening very much, rapidly using up its gas and dust to make clusters of stars and isolated stars and scattering them about. Such a galaxy would be elliptical in shape. A few billions of years after it had formed, there would be practically no interstellar gas and dust left, and practically no young stars.

A gas cloud with greater rotation might lead to the kind of spiral pattern we find in our own home galaxy. Five to ten billion years after birth, such a spiral galaxy still has much interstellar gas and dust in its disk, so star formation continues. Our galaxy and M31 are examples.

But no one yet knows how the irregular galaxies, like the Clouds of Magellan, fit into the picture. They are packed with young stars, but they also contain old objects such as globular clusters. They have some rotation but not very much. We still have a lot to learn about the life histories of galaxies.

A STEADY-STATE UNIVERSE?

What if the galaxies, like the stars, are not all the same age? Then some of them we now see out there are old, others are young, and still others are middle-aged. While some are dying and fading out of sight, others are being born out of supernebulae.

According to the *steady-state theory*, there is always enough hydrogen gas between the galaxies to form new galaxies. As ten, twenty, or a thousand galaxies fade from view, an equal number of new ones are formed and replenish the Universe. In this way, there is a more or less constant number of galaxies all the time, and the Universe always remains about the same as we see it now. But what happens to old, "dead" galaxies?

To date, no one has detected one. Another bothersome question related to the steady-state theory of the Universe is where the infinite supply of hydrogen for new galaxy formation comes from. So far, no one has turned up evidence for its source.

According to the founders of this theory—Thomas Gold, Fred Hoyle, and Herman Bondi—the creation of one hydrogen atom per year in a volume of space equal to that of the Empire State Building is enough to keep the Universe in a steady state. As the galaxies continue to rush away from each other and leave gaps of increasing size, new galaxies come into being and fill the gaps, say the steady-state theorists. So we must picture a Universe in which the density of matter remains constant while the volume increases. But where do the new hydrogen atoms required to keep the density constant come from?

According to Gold, they create themselves out of nothing! The steady-state theory violates a fundamental law of physics that has been around for a long time, the principle of the conservation of mass and energy. This law says that matter can neither be created nor destroyed. According to the astronomer Robert Jastrow, "It seems difficult to accept a theory that ignores such a firmly established fact of terrestrial experience."

THE BIG BANG

As we look around us in space, we see the galaxies rushing away from us in all directions. Have they always been doing so? Suppose for a moment that we could make time run backward. We would see the galaxies slow down, stop in their tracks, and then begin backing up, each one moving toward us at exactly its rushing-away speed. If we imagine the galaxies continuing to move in this way, what is bound to happen? They would retrace their paths and eventually come together in one region of space. The apparent outward rush of the galaxies at their observed speeds suggests that they were bunched up together some 12 to 20 billion years ago.

The *Big Bang theory*, first proposed in 1927 by the Belgian cosmologist Father Georges Lemaitre, envisions the Universe beginning as a super-dense "cosmic fireball" that exploded with tremendous force. Between 100,000 and 1 million years after the Big Bang, mammoth clouds of hydrogen and helium—the only two elements in the Universe at the time—formed. These clouds of gas became the matter out of which the protogalaxies formed. Star formation then began within each of the protogalaxies. It now seems that all of the galaxies formed at about the same time, during the first few billion years after the Universe was seeded with hydrogen gas. We have not observed any galaxies older than about 10 billion years, and we have not observed any new ones. But that does not necessarily mean that there aren't any.

Stars with large amounts of mass seem to have formed first, the stars that become supernovae. When supernovae explode, vast amounts of heavy elements are created and cast off, elements such as carbon, silicon, oxygen, iron, and others. The cast-off matter from early supernovae enriched the nebulae out of which Sunlike stars formed. Stars that contain an enrichment of heavy elements along with lots of hydrogen and dust are said to be *Population I stars*. These stars are found in open clusters of stars, such as the Pleiades, and in the spiral arms of spiral galaxies. *Population II stars* are older, redder, have relatively fewer metals, and are typical of globular clusters.

As a galaxy's matter is gradually used up in star formation, we can expect its character to change. Star formation slows and eventually stops. Meanwhile, the blue supergiant stars that light up the galaxy begin to flicker and die. Next, the yellowish Sunlike stars fade away to white dwarfs and eventually disappear from view as black dwarfs. Our galaxy is now populated mainly by the longer-lived red dwarf stars, so it shines dimly with a reddish light from its dwindling red giants and red dwarfs. Later still, a galaxy fades from view, becoming a dark, cold, and desolate place.

If wholesale stellar death signals the end of a galaxy, does it also signal the eventual end of the Universe? A few pages ago we left the Universe expanding at breakneck speed. Is it destined to keep on expanding forever? Or will something stop it?

THE BIG CRUNCH

If there is enough matter in the Universe, gravitational attraction will apply a braking action and slow down its expansion. The galaxies will stop the outward rush, pause briefly, and then slowly begin to tumble inward. Their matter will again fuse into a single cosmic fireball in what some astronomers call the "Big Crunch." Then there might be another Big Bang, and the Universe might be born anew. Once again galaxies might be hurled off in all directions. But eventually, the expansion would probably again be slowed, and another Big Crunch would occur. Astronomers call this model of the Universe an *oscillating Universe*—or the Bang, Bang, Bang theory.

One problem with an oscillating Universe, according to some astronomers, is that there may not be enough matter in the Universe to apply the gravitational brakes and slow down the expansion. We may be living in a Universe that is destined to just keep on expanding while its galaxies age and the stars flicker out, one by one.

Which view—if either—is correct, we cannot say. In their exploration of the Universe, astronomers and physicists are at the two great frontiers of science. Having smashed the atom into its many bits and pieces, physicists are now trying to discover patterns among those pieces. But the more they look, the more bits and pieces they find, and the more confusing the picture seems to become. It would seem that the fate of physicists is to learn more and more about less and less.

At the other extreme are the astronomers, who are trying to fit cosmic bits and pieces together and find patterns. But the more they find, the more confusing the picture becomes for them also. The fate of astronomers, it seems, is to learn less and less about more and more.

Both groups of scientists are exploring worlds that could not even be imagined before the beginning of this century. By the end of it, perhaps, we will understand a little more than we do now.

Glossary

ABSOLUTE ZERO. The temperature at which all molecular motion was presumed by Lord Kelvin to stop. On the Kelvin temperature scale, absolute zero is written as 0 K, or 0 kelvins, and is equal to $-273°C$.

ANGULAR DISTANCE. The distance between two objects measured by an angle. The angular distance between two stars, for example, could be found if you were to project a line to Star A and another line to Star B. The angle formed at your position would be the angular distance.

APPARENT BRIGHTNESS. The measure of a celestial object's brightness—how bright the object appears to the eye as opposed to its actual brightness, or luminosity. The farther away a light source is from the observer, the less bright it will appear, although its luminosity does not change.

APPARENT MOTION. The motion of any celestial object as seen from the Earth, which is itself moving.

ASTRONOMICAL UNIT. (a.u.) A measure of distance equal to the average distance of the Earth from the Sun, which was established by the U.S. Naval Observatory as 149,600,000 kilometers (92,760,000 miles).

ASTRONOMY. The science dealing with celestial bodies—their distances, luminosities, sizes, motions, relative positions, composition, and structure. The word comes from the Greek and means the "arrangement of the stars."

ATOM. The smallest possible unit of an element that can take part in a chemical reaction. An atom retains all the properties of its element. See also ELEMENT.

BINARY STARS. Two stars held in gravitational association with each other and revolving around a common center of mass. Also called *double stars*. Some star systems, such as the one to which Alpha Centauri belongs, have three or more stars held in gravitational association and are known as *multiple-star systems*.

BLACK DWARF. A star that has passed through the white dwarf stage and is radiating so little energy that it can no longer be observed directly.

BLACK HOLE. The incredibly dense remains of a massive star that has burned itself out. Black holes are thought tc be so dense that radiation is unable to escape from them.

BLUE GIANT. An especially massive, large, and luminous star, such as Rigel I, which is seen to shine with a bluish-white light. The core temperatures and surface temperatures of these short-lived stars are many times higher than those for less massive stars such as the Sun.

CELESTIAL EQUATOR. The equator on the celestial sphere; more properly, the circle formed where the plane of the Earth's equator cuts the celestial sphere.

CELESTIAL SPHERE. The stars and planets appear to move along the inner surface of a great hollow globe, one hemisphere of which you see as you observe the stars. The Earth appears to lie at the center of this imaginary sphere.

CEPHEID VARIABLE. A pulsating variable star with a brightness period of from two to about forty days. Because there is a relationship between the period and luminosity of Cepheid variables, these stars can be used to determine the distance to other galaxies.

CONSTELLATION. The grouping of certain stars or clusters of stars on the celestial sphere into imaginary figures. The ancients saw some of these groups of stars as human and animal figures, for example, Orion the Hunter. By international agreement, astronomers today recognize 88 constellations. Don't be discouraged, though, if you can't see a hunter, a bear, an archer, a lion, or any of the other imaginary figures in the sky. No one else can, either.

DEFERENT. According to Ptolemy, a planet revolves in a small circle, the center of which traces a larger circle with the Earth at its center. The larger circle is the deferent.

DENSITY. Mass per unit volume, or the amount of matter contained in a given volume of space, and expressed as grams per cubic centimeter. Water, for example, has a density of 1 g/cc.

ECLIPTIC. The path in the sky along which the Sun appears to travel in one year. This path forms a great circle on the celestial sphere.

ELEMENT. A substance made up entirely of the same kinds of atoms. Such a substance cannot be broken down into a simpler substance by chemical means. Examples are gold, oxygen, lead, and chlorine.

EQUINOXES. The two days of the year when the hours of sunlight and darkness are very nearly the same all over the Earth. This occurs around March 21 (vernal equinox) and September 21 (autumnal equinox).

FIRST POINT OF ARIES. That point where the Sun crosses the celestial equator at the vernal equinox. Although the vernal equinox occurred in Aries many centuries ago, it now occurs in the constellation Pisces.

GALAXY. A vast assemblage of stars, gas, and dust held together gravitationally in space. These "island universes" are classified according to their structure—for example, spiral, barred, elliptical, and irregular.

GEOCENTRIC SYSTEM. Until the 1500s, astronomers supposed that the Earth formed the center of the Universe and that all objects observed in the heavens revolved around the Earth (geo=Earth, centric=center).

GLOBULAR CLUSTER. A somewhat ball-like assemblage of 100,000 or more stars. A halo of about 100 globular clusters forms a sphere around the central part of our galaxy. Such clusters are also observed in certain other galaxies.

GLOBULES. Especially dense concentrations of gas and dust that appear to be the first stages of star formation. Globules have been identified in several large nebulae.

GRAVITATION. The force of attraction that exists between any two or more objects in the Universe, no matter how large, small, or far apart the objects are. This attraction is proportional to the mass of the objects involved and inversely proportional to the square of the distance between them. The greater the mass, the greater the force of attraction; the greater the distance, the less the force of attraction.

HELIOCENTRIC SYSTEM. After the 1500s, astronomers came to recognize that the Sun formed the center of the Solar System and that all of the planets revolved around the Sun (helio=Sun, centric=center).

INVERSE SQUARE LAW OF LIGHT. A light source appears to diminish or brighten, depending on our distance from it. The light intensity will appear inversely proportional to the square of our distance from it; for example, double the distance and the intensity appears four times fainter.

LIGHT-YEAR. The distance that light travels in one year. Since it moves at the rate of about 300,000 kilometers a second, this amounts to about 10 trillion kilometers.

LOCAL GROUP. The group of about twenty galaxies to which our home galaxy, the Milky Way, belongs. The volume of space containing the Local Group has a diameter of about 3 million light-years.

LUMINOSITY. The total amount of radiation emitted by an object.

MASS. A given quantity of matter of any kind.

MILKY WAY. The name of our local galaxy, containing some 300 billion or more stars. Also the name of that hazy band of light seen in the summer and winter sky. A small telescope resolves the band into individual stars.

MOLECULE. Two or more atoms of the same or different elements chemically combined. The smallest unit of a substance above the size of an individual atom of that substance. The smallest quantity of water, for example, is a water molecule consisting of one atom of oxygen and two atoms of hydrogen.

NEBULA. A great cloud of gas and dust within a galaxy. Some nebulae, called *reflection nebulae*, reflect light generated by nearby stars, or by stars embedded within the nebula. Other nebulae are dark. Still others reradiate energy emitted by stars embedded in the nebulae; these are called *emission nebulae*. And still others take the form of a great shell of gas cast off by an eruptive or explosive star. These are called *planetary nebulae* partly because they were once mistaken for planets within the Solar System.

NEUTRON. An electrically neutral particle in the nucleus of atoms. All atoms except hydrogen contain neutrons. Neutrons are only slightly more massive than protons. Outside the atom, neutrons have a life of only twenty minutes or so before decomposing into electrons and protons and giving off gamma rays.

NEUTRON STAR. A star made up of neutrons. Because neutrons are without an electrical charge and there is no force of repulsion, they can be packed very closely together. Consequently, neutron stars are extremely dense objects.

NOVA. A star that, for some reason not yet fully understood, bursts into brilliance. Within a few days a typical nova may become hundreds of thousands of times brighter than usual, then somewhat less brilliant. After a few months or so the star returns to its pre-nova brightness. Certain planetary nebulae may be the result of nova eruptions.

NUCLEAR FUSION. The union of atomic nuclei, and, as a result, the building of nuclei of more massive atoms. Hydrogen nuclei in the core of the Sun fuse and become the nuclei of helium atoms. In the process, large amounts of energy are released.

NUCLEUS. In astronomy, the central portion of a galaxy or of a comet. In chemistry and physics, the central portion of an atom.

ORBIT. The path one celestial object traces as it moves around another to which it is attracted by the force of gravitation. The Earth and the other planets of the Solar System all have their own orbits around the Sun. The Moon travels in an orbit around the Earth, and the Sun travels in an orbit around the nucleus of the Milky Way Galaxy.

PARALLAX. The apparent change in position of an object when it is viewed from two different positions. The object appears to change place against the background of more distant objects. By viewing a nearby star from opposite points in the Earth's orbit, astronomers can measure the star's angle of apparent shift, the parallax, and so determine the star's distance from us.

PERIOD. The time a variable star takes to complete one cycle of going from bright to dim and back to bright again. The periods of some variables are measured in hours, while others are measured in weeks or months. Also, the length of time it takes one celestial object to complete one orbit around another.

PHOTON. A quantum, or particle, of light. These packets of energy are characteristic of particular wavelengths of light, so we can speak of violet photons being more energetic than red photons.

PLANET. A celestial object that shines by reflected light from a star about which the planet is held gravitationally captive and revolves. There are nine known primary planets in the Solar System.

PLANETARY SYSTEM. Any star accompanied by one or more planets. The Solar System is presumably one of numerous planetary systems in the Milky Way Galaxy.

POPULATION I STAR. A star, like the Sun, that is associated with clouds of interstellar dust and gas in the spiral arms of the Milky Way Galaxy. Population I stars are relatively young and are formed at least partly from material cast off by supernovae.

POPULATION II STAR. A type of star found in the dense nucleus of our galaxy; usually part of a globular cluster. Population II stars are not associated with interstellar dust, and so are not believed to be forming now. They are old stars.

PRECESSION. A gradual change in the direction of tilt of the Earth's axis, due to gravitational attraction of the Sun and Moon, which tend to pull the Earth's equatorial bulge into line. This double attraction causes the Earth to wobble slightly, like a spinning top. The axis completes one rotation about every 26,000 years, which means that Polaris has not always been the North Star, nor will it continue to be.

PROTOGALAXY. A galaxy in the process of formation out of a cloud of gas and dust.

PROTOSTAR. A star in the process of formation out of a cloud of gas and dust.

PULSAR. A rapidly rotating neutron star that sends out "pulses" of radiation. When the radiation is emitted in the direction of the Earth, we receive a pulse. About 150 or so pulsars have been detected; each has its own rate of pulsation.

RED DWARF. A star with relatively little mass and a low surface temperature (about 3,000 kelvins), which causes the star to shine with a reddish light.

RED GIANT. An enormous star that shines with a red light because of its relatively low surface temperature (about 3,000 kelvins). It is now thought that most stars—except red dwarfs—go through a red-giant stage after they exhaust their core hydrogen and the core collapses inward. The star then swells up, becoming a red giant.

REVOLUTION. The motion of one body around another. The Moon revolves around the Earth; the planets revolve around the Sun.

ROTATION. The motion of a body around its axis. The Sun and all of the planets rotate, with the Earth completing one rotation about every twenty-four hours.

RR LYRAE VARIABLE STAR. A pulsating variable star with a period of less than one day. Because the periods of these stars are closely associated with their average luminosities, the stars can be used as distance indicators.

SOLAR SYSTEM. The Sun, its nine known planets accompanied by about 45 moons, plus many lesser objects, including comets, asteroids, meteoroids, and one planetoid (Chiron).

SOLSTICES. The highest and lowest points from the celestial equator reached by the Sun as it appears to travel along the ecliptic. The northernmost point is called the *summer solstice* (about June 22). The southernmost point is called the *winter solstice* (about December 22). The summer solstice point lies in Gemini and the winter solstice point lies in Sagittarius. On arriving at these points, the Sun appears to stand still momentarily.

SPECTROSCOPE. A *prism spectroscope* is an instrument fitted with a prism that separates a star's light into its individual colors, or spectrum. A *diffraction grating spectroscope* is a spectroscope fitted not with a prism but with a polished glass surface containing many thousands of grooves per centimeter. When a star's light falls on this grating, the different colors making up the light are diffracted at different angles and so form a spectrum. A diffraction grating spectrum of a star is sharper than a prism spectrum.

STAR. A hot, glowing globe of gas under great pressure that emits energy. The Sun is a typical, and our closest, star. Most stars are enormous compared with planets, containing enough matter to make thousands of Earthlike planets. Stars generate energy by the fusion of atomic nuclei in their dense, hot cores. Sunlike stars seem to be formed out of dense clouds of gas and dust, evolve through various stages, and finally end their lives as dark and cold objects called *black dwarfs.*

SUPERNOVA. A giant star whose brightness is tremendously increased by a catastrophic explosion. Supernova stars are many thousands of times brighter than nova stars. In a single second, a supernova releases as much energy as the Sun does over a period of about sixty years.

TEMPERATURE. A measure of how hot or cold a body is, "hotness" meaning the rate of atomic motion, or kinetic energy. The greater the kinetic energy, the "hotter" a substance is said to be.

VARIABLE STAR. A star whose energy output varies, either regularly or irregularly.

VOLUME. A given amount of space.

WAVELENGTH. The distance between two successive crests or troughs of a wave of any kind. The wavelength is found by dividing the velocity of a wave by its frequency.

WHITE DWARF. A very small star that radiates stored energy rather than new energy generated through nuclear fusion. The Sun is destined to become a white dwarf after it goes through the red-giant stage.

ZODIAC. The zone 16 degrees wide stretching around the sky and centered on the ecliptic. The Moon, Sun, and most of the planets move about the sky in this zone, at least as seen from the Earth.

Index

Absolute scale, 41
Absolute zero, 41, 77
Alexandria, 10
Alpha Centauri, 40
Andromeda Galaxy, 64–67
Angle of parallax shift, 39
Angular distance, 37, 77
Apollonius, 19
Apparent brightness, 6, 77
Apparent motion, 77
Aquarius constellation, 15, 58, 59
Aratus of Soli, 6
Aries, 15
Aristarchus, 12
Aristotle, 10
Artificial satellites (table), 32
Astronomical unit (a.u.), 39, 77
Astronomy, 77
Atom, 77
Auriga, 47
Autumnal equinox, 15

Bang, Bang, Bang Theory, 76
Barnard's Star, 39–40
Barred spiral galaxy, 69
Bellatrix, 6
Berenice's Hair, 8
Bessel, Friedrich, 37
Betelgeuse, 6, 40
Big Bang Theory, 75–76
Big Crunch, 76
Binary stars, 77
Black dwarf, 53, 75, 78
Black hole, 53, 55, 61, 78
Blue giant, 78

Blue-white supergiant stars, 55
Bok, Bart, 63
Bondi, Herman, 74
Bootes constellation, 73
Bruno, Giordano, 25–26

Calculus, 30
Cancer, 15
Capella, 40
Capricorn, 15
Carina arms of galaxy, 55
Castor, 35
Catalogue of the Southern Stars
   Halley, 33
Celestial equator, 14, 78
Celestial sphere, 6, 78
Celsius scale, 41
Cepheid variables, 65, 66, 78
Closed spiral system, 69
Clouds of Magellan, 62–63, 66, 69, 74
Columba constellation, 35
Constellations, 5–6, 15, 78
Copernican Revolution, 22
Copernicus, 22, 25–26
Corona of Milky Way, 61–63
Cosmic year, 63
Crab Nebula, 60, 61
Cygnus the Swan, 37, 60
Cygnus X-I, 55

Dark nebula, 58
Dead galaxy, 74
Deferent, 19, 78
Density, 78

Digges, Thomas, 25, 33
Distance to stars, measuring, 38–40
    to distant stars, 45–46
Double stars, 35, 43–44
Dwarf galaxy, 66
Dying stars, 58

Earth
    circumference of, 11
    shape of, 10, 16
    size of, 10
Ecliptic, 12, 78
Eddington, Arthur, 43–45
Einasto, J., 62
Electromagnetic spectrum, 52
Element, 78
Ellipse, 27, 28
Elliptical galaxies, 69, 70
Emission nebulae, 58
Epicycles, 19-20, 22, 23, 25
Equinoxes, 15–17, 79
Eratosthenes, 10–11
Eta Aurigae, 40
Eudoxus, 8–9, 25
Event horizon, 53

Fahrenheit scale, 41
First Point of Aries, 16, 79
Fixed stars, 8
Foci, 27

Galactic corona, 61–63
Galactic disk, 55–58
Galactic nucleus, 51–53, 55
Galactic year, 63
Galaxies
    Andromeda, 64, 65, 67
    barred spiral, 69
    center of, 50
    dead, 74
    discovery of, 66
    dwarf, 66
    elliptical, 69, 70
    evolution of, 74
    irregular, 69
    M, 69
    movement of, 70–71
    nucleus of, 51–55, 80
    open-spiral, 69
    peculiar, 69
    Sombraro, 51
    speed of, 70–71
    spherical, 69
    spiral, 33, 55, 68, 70, 74
    spiral arms of, 55–58
    superclusters, 66
    transitional, 69
    types of, 69–70
Galileo, 29–30, 64
Gamow, George, 2
Gemini, 15
Geocentric system, 22, 79
Globular clusters, 49, 50, 62, 69, 79
Globules, 79
Gold, Thomas, 74, 75
Gravitation, 16, 30–31, 62, 79
    tug on earth, 17
Great Nebula in Andromeda, 65, 66
Great Nebula in Orion, 58

Halley, Edmund, 33
Heat of stars, 40–41
Heliocentric system, 22, 79
Hercules constellation, 35, 64, 66
Herschel, William, 34–36, 65
Hertzsprung, Ejnar, 42, 43
Hipparchus, 11–12, 16
Horsehead Nebula, 57–58
House of Wisdom, 21
Hoyle, Fred, 74
Hubble, Edwin, 65, 66
Hydrogen atoms, 74, 75

Interstellar hydrogen, 70
Inverse square law of light, 45, 79
Irregular galaxies, 69

Jastrow, Robert, 75

Kelvin scale, 40–41
Kepler's Laws, 27–28
Koppernigk, Nicholas. *See* Copernicus

Large Cloud of Magellan, 62–63
Law of gravity, 16, 30–31, 62, 79
Laws of motion, 30–32
Lemaitre, Georges, 75
Leo, 15
Libra, 15
Light year, 37, 79
Local group, 66, 79
Luminosity of stars, 6, 42–43, 80

M galaxies, 69
M objects, 64
Magellan, Clouds of, 62–63, 66, 69, 74
Main sequence, 43
Mass, 80
Mass-luminsoity, 43–44
Mass of stars, 43–44
Messier, Charles, 64
Milky Way, 80
    corona of, 61–63
Mirach, 64
Molecule, 80
Moon, size of, 11–14
Motion, laws of, 30–32
Mount Wilson Observatory, 65

Nebula, 80
Neutron, 80
Neutron star, 6, 80
Newton, Isaac, 16, 30–32
NGC 1300, 69
North Star, 17
Nova, 61, 80
Nuclear fusion, 80
Nucleus of galaxy, 51–55, 80
Nun, 5

"Objects to avoid," 64
On the Revolutions of the Heavenly Spheres (Copernicus), 22
Open spiral galaxy, 69
Optical illusions, 25
Orbit, 81
Orbital period, 31
Origin and future of the Universe, 73–76
Orion arm of the galaxy, 55

Orion the Hunter, 6, 7, 58, 59
Oscillating Universe theory, 76

Parallax, 81
Parallax angle, 38, 39
Parallax method, 37
Parallax shift, 37, 38, 39
Peculiar galaxies, 69
Period of star, 50, 81
Perseus arm of galaxy, 55
Photon, 81
Pisces, 15
Planetary nebulae, 52, 58
Planetary system, 81
Planets, 8, 81
    speed around sun, 62
Pleiades, 58
Polaris, 17
Polestar, 17
Population stars (1 and 2), 68, 75, 81
Precession of equinoxes, 16, 17, 81
Principia (Newton), 32
Protogalaxy, 82
Protostar, 82
Proxima Centauri, 40
Ptolemaeus, Claudius (Ptolemy), 19–20, 21
Pulsars, 61, 82
Pythagoras, 10

Quasars, 72

Radio telescopes, 58, 70
Red dwarf, 43, 82
Red-giant stars, 51–53, 82
Reflection nebulae, 58
Revolution, 82
Rigel, 6
Ross 248, 43
Rotation, 82
    of clouds, 74
Royal Observatory at Greenwich, 34
RR Lyrae variable stars, 50, 52, 69, 82
Russell, Henry Norris, 42–43

Sagittarius, 15, 47, 51
Sagittarius arm of galaxy, 55
Scorpio, 15
Shapley, Harlow, 36, 47, 50
Singularity, 53
Sirius, 40
61 Cygni, 37, 40
Size of galaxy, 47–50
Small Cloud of Magellan, 63
Solar System, 55, 82
Solstices, 14, 82
Sombraro Galaxy, 51
Space Telescope, 58
Spectroscope, 73, 82–83
Speed of galaxies, 70–71
Speed of light, 71
Spherical galaxies, 69
Spica, 40, 43
Spiral arms of galaxy, 55–58
Spiral galaxies, 33, 55, 68, 70, 74
Spirals, 69
Star gauges, 36
Starry Messenger, The (Galileo), 29
Stars, 83
    binary, 77
    blue-white supergiant, 55
    color of, 40, 42–43
    distance to, 38–40, 45–46
    double, 35, 43–44
    dying, 58
    fixed, 8
    formation, 53, 55, 75
    gauges, 36
    heat of, 40
    hottest, 58
    luminosity of, 6, 42–43, 80
    mass of, 43–44
    movement of, 33
    neutron, 61, 80
    North, 17
    period of, 50, 81
    pole–, 17
    population 1 and 2, 68, 75, 81
    proto–, 82
    red giant, 51–53, 82
    temperature of, 42–43
    variable, 50, 83
    wandering, 8–9, 12
Steady-state theory, 74–75
Sun, motion of, 35
Supercluster of galaxies, 66
Supernova, 26, 61, 75, 83
Syene, 10

Taurus, 15, 60, 61
Telescope, use of, 29
    radio, 58, 70
    Space, 58
Temperature, 83
    of stars, 42–43
Temperature degrees, 41
Temperature–Luminosity, 42
Thompson, William, 41
3C 295, 71
Transitional galaxies, 69
Trumpler, Robert, 50
"Tycho's Star," 26

Uraniburg Observatory, 26
Uranus, 34

Variable stars, 50, 83
Veil Nebula, 60, 61
Vernal equinox, 15
Virgo, 15
Volume, 83

Wandering stars, 8–9, 12
Wavelength, 83
White dwarf, 52, 53, 75, 83

Zeta Canis Majoris, 40
Zodiac, 12–15, 83

About the Author

Roy A. Gallant is adjunct full professor of English at the University of Southern Maine and director of the university's Southworth Planetarium, where he creates planetarium shows and lectures to student and adult groups. When his schedule permits, he teaches a graduate level and undergraduate course in nonfiction writing. He is former editor-in-chief of the Natural History Press, of the American Museum of Natural History in New York City; executive editor of Aldus Books, Ltd, of London (a subsidiary of Doubleday); and managing editor of *Scholastic Teacher* magazine. He also was a member of the faculty at the American Museum-Hayden Planetarium in New York from 1972 to 1978 and taught astronomy there to adults and gifted students. He served as a science education specialist for the University of Illinois Elementary Science Study Program. Currently he is an earth science consultant for the magazine *Science and Children*, published by the National Science Teachers Association, a fellow of the Royal Astronomical Society of London, and member of the New York Academy of Sciences.

Professor Gallant is the author of more than fifty science text and trade books for young readers and adults, many in the area of astronomy. Among his most recent works are a paper on the creation/evolution controversy (Oxford University Press, 1983) and the books *The Planets: Exploration of the Solar System* (Four Winds Press, 1982) plus the distinguished *Our Universe* (National Geographic Society, 1980).